Mastering jBPM6

Design, build, and deploy business process-centric
applications using the cutting-edge jBPM
technology stack

Simone Fiorini

Arun V Gopalakrishnan

[PACKT] open source *
PUBLISHING   community experience distilled

BIRMINGHAM - MUMBAI

Mastering jBPM6

First published: June 2015

Production reference: 1220615

Published by Packt Publishing Ltd.
Livery Place
35 Livery Street
Birmingham B3 2PB, UK.

ISBN 978-1-78328-957-8

www.packtpub.com

Credits

Authors
Simone Fiorini
Arun V Gopalakrishnan

Reviewers
Ruben Andres Ghio
Jerónimo Ginzburg
Federico Weisse

Commissioning Editor
Rukhsana K

Acquisition Editor
Vinay Argekar

Content Development Editors
Anand Singh
Sweny M. Sukumaran

Technical Editor
Tanmayee Patil

Copy Editor
Tani Kothari

Project Coordinator
Vijay Kushlani

Proofreader
Safis Editing

Indexer
Rekha Nair

Graphics
Valentina D'silva
Jason Monteiro

Production Coordinator
Aparna Bhagat

Cover Work
Aparna Bhagat

About the Authors

Simone Fiorini is a seasoned professional with 20 years of software development experience, specializing in large-scale software systems, mission critical applications, and project management in a wide variety of computing environments and platforms, with a focus on EAI, BPM, and integration-oriented tools. His latest efforts are focused on an online reservation system for a large Middle East railway company and a scalable, reactive, financial market data server for a leading Italian banking group's investment bank.

A graduate of both Università di Parma (earth science) and University of Milan (engineering of computing systems), Simone resides near Brescia, where he's trying to grow roses and other fragrant flowers for his wife, Giuliana, and their two sons.

> I'd like to thank my wife, Giuliana, and my sons as they support everything I do any way I do it, my parents, my brothers, and all the people who made this book possible.

Arun V Gopalakrishnan has more than 9 years of experience in creating architecture, designing, and developing enterprise applications used in the domains of BFSI, supply chain management, and telecom businesses.

He graduated as a bachelor of technology in electronics and communications and holds a master's degree in software systems. Currently, he is playing the role of a software architect in designing and developing an enterprise middleware platform.

He is well versed in service-oriented, event-driven architectures and has experience in integrating jBPM with enterprise architecture. He is passionate about learning new technologies and improving his knowledge base and expertise in JEE, OOAD, SOA, EDA, expert systems, and distributed computing.

Along with this technical expertise, he enjoys engineering software applications and is an avid follower, practitioner, and mentor of agile and continuous delivery models. He believes in tools and is well versed in creating, using, and extending DevOps tools, such as Eclipse, Maven, Git, Gerrit, Jenkins, Sonar, JIRA, and MediaWiki, to enhance the efficiency of development teams.

First and foremost, I would like to thank the editors Vinay Argekar, Sweny Sukumaran, Tanmayee Patil, and Anand Singh for their immense support and help in getting this book ready for publishing. They made the process smooth and enjoyable and were patient and cheerful with those painful follow-ups when I missed deadlines. I would like to take this opportunity to thank the technical reviewers, Federico Weisse, Jerónimo Ginzburg, and Ruben Andres Ghio, for their feedback, which helped shape this book.

Authoring a book with an already tight regular job always calls for long weeks. Without my supportive family, I could not have taken yet another step in my career.

My mentors and colleagues were always my source of confidence and energy. I would like to thank them for the knowledge, experience, and goodwill that they share with me. It has always been great working with you.

Sincere gratitude goes to the open source community, particularly the jBPM community, which has supported me a lot in understanding technologies and tools. I am grateful for working with them and hope that their members will find this book useful!

To the Almighty, for surrounding me with these beautiful people and enjoyable opportunities.

About the Reviewers

Ruben Andres Ghio was born on June 5, 1981, under the zodiac sign of Gemini in the city of Quilmes, Argentina. When he was 2 years old, he moved to Villa Gesell, a small city over the sea. This place was his home for almost 11 years. After that, he returned to Buenos Aires, where he got a high school degree with a specialization in economics.

At the age of 22, he graduated as a system engineer from the National Technological University of Buenos Aires.

As a River Plate fan, Ruben has practiced a lot of different sports, but he has always had special feelings for his two favorites: tennis and soccer, which he still practices. You can follow him on Twitter at @rubenghio.

Ruben has been working with technology since 2002, starting as a Visual Basic developer for one of the most important banks in Argentina. After 2.5 years, he moved to one of the biggest technology companies in the world, where he started working as a Java developer, and after almost 3 years, he started to focus on his career as an IT architect, a skill he developed for the next 4 years. Currently, he is working as a senior IT architect at one of the biggest open source companies in the world, developing and building modern and innovative projects for government and private companies across Latin America.

I would like to thank my wife, who always supports me in all the things I do, suggesting, encouraging, and helping me out.

Jerónimo Ginzburg works at Red Hat as a senior architect in the consulting team. He has a computer science degree from the University of Buenos Aires (UBA) and over 15 years of experience in developing Java Enterprise applications. He is a strong believer in open source and agile software development. Jerónimo joined Red Hat in 2008 and, since then, has helped many Latin American companies adopt Drools and jBPM. He has co-written several software engineering articles published in refereed conferences and journals, and he was also the technical reviewer for the book *jBPM Developer Guide*, *Packt Publishing*.

I would like to thank my beautiful wife, Mariana, and my newborn son, Ramiro, for their love, patience, and support during this revision.

Federico Weisse was born in Buenos Aires, Argentina. He has over 15 years of expertise in the IT industry. In his career, he has worked with several technologies and programming languages, such as C, C++, ASP, and PHP; different relational databases (Oracle, SQL Server, DB2, and PostgreSQL); platforms (AS400, Unix, and Linux); and mainframe technologies.

In 2002, he adopted Java/J2EE as his main technology. He has been working with it since then and has become a specialist in this field. A couple of years later, he got involved with BPM systems and has been working with JBPM since 2009.

Nowadays, he is a J2EE architect at one of the most important healthcare providers in Argentina.

www.PacktPub.com

Support files, eBooks, discount offers, and more

For support files and downloads related to your book, please visit www.PacktPub.com.

Did you know that Packt offers eBook versions of every book published, with PDF and ePub files available? You can upgrade to the eBook version at www.PacktPub.com and as a print book customer, you are entitled to a discount on the eBook copy. Get in touch with us at service@packtpub.com for more details.

At www.PacktPub.com, you can also read a collection of free technical articles, sign up for a range of free newsletters and receive exclusive discounts and offers on Packt books and eBooks.

![PACKTLIB logo]

https://www2.packtpub.com/books/subscription/packtlib

Do you need instant solutions to your IT questions? PacktLib is Packt's online digital book library. Here, you can search, access, and read Packt's entire library of books.

Why subscribe?

- Fully searchable across every book published by Packt
- Copy and paste, print, and bookmark content
- On demand and accessible via a web browser

Free access for Packt account holders

If you have an account with Packt at www.PacktPub.com, you can use this to access PacktLib today and view 9 entirely free books. Simply use your login credentials for immediate access.

Table of Contents

Preface

jBPM is a leading open source BPM and workflow platform whose development is sponsored by Red Hat under Apache Software License (ASL) licensing. The jBPM product has been around for almost 10 years; its strongest points rely on flexibility, extensibility, and lightness, and it is a modular, cross-platform pure Java engine that is BPMN2 compliant.

It features a robust management console and development tools that support the user during the business process life cycle: development, deployment, and versioning. It integrates with widely-adopted frameworks and technologies (SOAP, REST, Spring, Java EE CDI, and OSGi) and provides off-the-shelf support to Git and Maven.

It fits into different system architectures and can be deployed as a full-fledged web application or as a service; it can be tightly embedded into a classical desktop application or loosely integrated into a complex event-driven architecture. In its default configuration, jBPM can be hosted by the enterprise class application server Red Hat EAP 6.x or the bleeding-edge Red Hat WildFly 8 server.

Mastering JBPM6 takes you through a practical approach to using and extending jBPM 6.2. This book provides a detailed jBPM 6.2 overview; it covers the BPM notation supported by the engine and explains the advanced engine and API topics focusing, as much as possible, on several working practical examples.

The book presents the user with solutions to common real-time issues like BAM (which stands for business activity monitoring) and production scenarios.

What this book covers

Chapter 1, Business Process Modeling – Bridging Business and Technology, gives the user an overview of the BPM environment, introduces the jBPM world and give insight to the big picture of business logic integrated platform.

Chapter 2, Building Your First BPM Application, starts by taking the user straight to the jBPM tool stack by providing the reader with a hands-on product installation and configuration tutorial, and then, it tackles beginner topics such as business process modeling and deployment.

Chapter 3, Working with the Process Designer, digs deep into web-based jBPM tools to illustrate to the user the main jBPM web designer features: user forms, scripting, and process simulation.

Chapter 4, Operation Management, describes the new jBPM artifacts architecture, focusing on Maven repositories (modules and deployment), engine auditing and logging analysis, jobs scheduling, and a full working BAM customization example (with Dashboard integration).

Chapter 5, BPMN Constructs, illustrates the BPMN2 constructs implemented by jBPM and provides insights and caveats about their usage by commenting a contextually ready-to-use source code example.

Chapter 6, Core Architecture, covers all the jBPM modules (for example, human task service, persistence, auditing, and configuration) by elaborating on how to leverage engine functionalities with the help of several source code examples.

Chapter 7, Customizing and Extending jBPM, explores engine customization areas with a practical approach; it provides the user with explanations on how to customize persistence, human task service, marshalling mechanism and the work item handler architecture.

Chapter 8, Integrating jBPM with Enterprise Architecture, describes how jBPM can integrate with external applications through SOAP, REST, or JMS either as a client or a server. It offers insights on how to leverage its services in a Java EE application.

Chapter 9, jBPM in Production, explores the jBPM system features when dealing with service availability, scalability, and security; it provides tips and techniques related to engine performance tuning in production environments.

Appendix A, The Future, briefly details the trends and future of Business Process Modeling.

Appendix B, jBPM BPMN Constructs Reference, is a quick reference for the BPMN constructs supported by jBPM.

What you need for this book

You will need the following software to be installed before running the code examples:

jBPM requires JDK 6 or a higher version. JDK 6 or newer versions can be downloaded from `http://www.oracle.com/technetwork/java/javase/downloads/index.html`. There are installation instructions on this page as well. To verify that your installation was successful, run `java -version` on the command line.

Download `jbpm-6.2.0.Final-installer-full.zip` from `http://sourceforge.net/projects/jbpm/files/jBPM%206/jbpm-6.2.0.Final/`. Just unzip it in a folder of your choice. The user guide (`http://docs.jboss.org/jbpm/v6.2/userguide/jBPMInstaller.html`) includes instructions on how to get started in a simple and quick manner.

The jBPM setup requires Ant 1.7 or later (`http://ant.apache.org/srcdownload.cgi`).

The additional required software is as follows:

- Git 1.9 or later (`http://git-scm.com/downloads`)
- Maven 3.2.3 or later (`http://maven.apache.org/download.cgi`)

The preferred development IDE to run the examples is the Eclipse Kepler distribution, which can be automatically downloaded and pre-configured with the BPMN installation process.

Who this book is for

This book is primarily intended for jBPM developers, business analysts, and process modelers, and, to some extent, for project managers who must be exposed to the jBPM platform features. The book assumes that you have prior knowledge of business analysis and modeling, and, of course, Java; basic knowledge of jBPM is also required.

Conventions

In this book, you will find a number of styles of text that distinguish between different kinds of information. Here are some examples of these styles, and an explanation of their meaning.

Code words in the text are shown as follows: "Specify a role for the users in the `roles.properties` file."

A block of code is set as follows:

```
ReleaseId newReleaseId = ks.newReleaseId("com.packt.masterjbpm6",
"pizzadelivery", "1.1-SNAPSHOT");
// then create the container to load the existing module
Results result = ks.updateToVersion (newReleaseId);
```

When we wish to draw your attention to a particular part of a code block, the relevant lines or items are set in bold:

```
<bpmn2:scriptTask id="_2" name="prepare order" scriptFormat="http://
www.java.com/java">
```

Any command-line input or output is written as follows:

```
ant install.demo
```

New terms and **important words** are shown in bold. Words that you see on the screen, in menus or dialog boxes for example, appear in the text like this: "Select the **Administration | Data Providers** link from the left navigation menu pane."

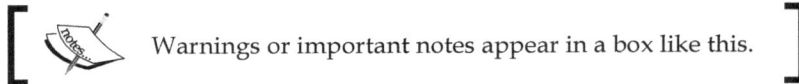

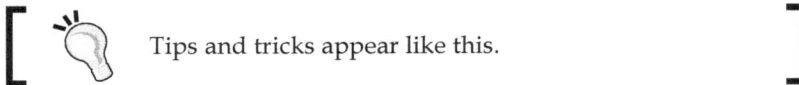

> Warnings or important notes appear in a box like this.

> Tips and tricks appear like this.

Reader feedback

Feedback from our readers is always welcome. Let us know what you think about this book—what you liked or may have disliked. Reader feedback is important for us to develop titles that you really get the most out of.

To send us general feedback, simply send an e-mail to feedback@packtpub.com, and mention the book title via the subject of your message.

If there is a topic that you have expertise in and you are interested in either writing or contributing to a book, see our author guide on www.packtpub.com/authors.

Customer support

Now that you are the proud owner of a Packt book, we have a number of things to help you to get the most from your purchase.

Downloading the example code

You can download the example code files from your account at `http://www.packtpub.com` for all the Packt Publishing books you have purchased. If you purchased this book elsewhere, you can visit `http://www.packtpub.com/support` and register to have the files e-mailed directly to you.

Errata

Although we have taken every care to ensure the accuracy of our content, mistakes do happen. If you find a mistake in one of our books—maybe a mistake in the text or the code—we would be grateful if you could report this to us. By doing so, you can save other readers from frustration and help us improve subsequent versions of this book. If you find any errata, please report them by visiting `http://www.packtpub.com/submit-errata`, selecting your book, clicking on the **Errata Submission Form** link, and entering the details of your errata. Once your errata are verified, your submission will be accepted and the errata will be uploaded to our website or added to any list of existing errata under the Errata section of that title.

To view the previously submitted errata, go to `https://www.packtpub.com/books/content/support` and enter the name of the book in the search field. The required information will appear under the **Errata** section.

Piracy

Piracy of copyright material on the Internet is an ongoing problem across all media. At Packt, we take the protection of our copyright and licenses very seriously. If you come across any illegal copies of our works, in any form, on the Internet, please provide us with the location address or website name immediately so that we can pursue a remedy.

Please contact us at `copyright@packtpub.com` with a link to the suspected pirated material.

We appreciate your help in protecting our authors, and our ability to bring you valuable content.

Questions

You can contact us at `questions@packtpub.com` if you are having a problem with any aspect of the book, and we will do our best to address it.

1
Business Process Modeling – Bridging Business and Technology

"All business does IT" (Information Technology). This was a superlative but a futuristically interesting tweet I came across recently. The impact of information technology on business in recent years has been overwhelming. They are like two large galaxies in our universe, colliding and merging. The current state of this merger can be defined by one term - collaboration. Businesspeople collaborate with information technology (IT) people and use IT services to continuously improve and deliver commercially viable and profitable product/services to their customers.

Collaboration quintessentially needs effective communication, and **Business Process Modeling** smoothly fits into this scenario. Business process modeling is not new; business people have always used it. Models were developed in mind, and then, they were written down as text or depicted as diagrams. With IT embracing business, these models evolved into standard flow charts and activity diagrams. However, there was ambiguity; the diagrams and text provided by business as requirements were interpreted by the technical people and they had their own representation, architectural models, requirement documents, and design. This was duplication of effort and people, intensive, often with a line of meetings between business stakeholders and information technologists negotiating, and arriving at conclusions about the business requirements. This being the scenario, what improvement does business process modeling bring to the table? Business process modeling brings in the concept of a common artifact between business and information technologists.

Business models are prepared by business analysts and shared with technologists. They collaborate and improve on the model and arrive at an artifact that is executable. Further, technologists as always reduce their work by automating their involvement; that is, we are moving largely to software systems where business people can configure the business process and execute it without the intervention of an information technologist.

Concluding the philosophy, let us jump to the larger context of **Business Process Management** (**BPM**) and discuss it in detail. This chapter covers the following concepts:

- Business process management concepts
- The standard – business process model and notation (BPMN 2.0)
- Use cases of BPM as applied in the industry
- Design patterns in the BPM world
- A brief introduction to jBPM
- Business logic integration platform, the bigger picture

Business process management

BPM involves the designing, modeling, executing, monitoring, and optimizing of business processes. A specialized software system that helps achieve these objectives completely is called **Business Process Management System** (**BPMS**). Most of the IT infrastructure used in business is in fact part of one or more business processes, and a BPMS should have the capability to manage the complete life cycle of a business management process. Further, jBPM provides a complete BPMS.

A business process consists of a set of activities, organized to complete a specific business objective, which varies from *creating a product* to *delivering a service*. A business process model also provides a visual representation of the business processes. The activities in a business process (also called tasks) are connected to represent the execution flow of a business process; further, these activities can be categorized.

jBPM helps its users to define and model business processes by using its process designers. Business users can in fact design the business process online, and efficient versioning and history capabilities help in making the activity of modeling business processes collaborative. There are provisions to simulate how a business process might behave during runtime. jBPM also provides capabilities to migrate the process definition to an updated version when business changes stipulate process improvements.

The defined business process models are deployed to BPM software, where instances of these process definitions are created to execute the process. jBPM provides the capability of executing business processes and has complete operation management capabilities such as tracking, controlling, and maintaining the history of the life cycle of all process instances.

Human interaction management

An important concept that we need to discuss in detail while explaining BPMS is **Human Interaction Management (HIM)**. The activities that form business processes can be broadly classified into **automatic** and **manual**. Automatic activities are those that can be completed by the software system without any manual intervention. For example, in a banking transaction process, the customer has to provide the details of the transaction, such as the bank account number to which the money needs to be transferred and the amount to be credited into this account. This is a manual activity. In contrast, if an SMS alert needs to be sent to the customer's mobile as part of this process, no manual intervention is needed; the software can fetch the mobile number registered with the customer's account and send the SMS automatically. An SMS alert is an automatic activity in the banking transaction process.

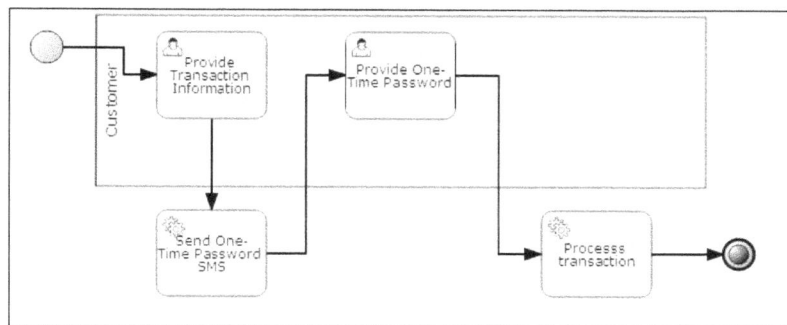

Therefore, human tasks are in fact human interactions in the business processes or, in a broader sense, human interactions with the software system in place. Usually, human activities are physical tasks that are performed outside the software system and the results or conclusions are then fed as input into the software system. Technically, from the perspective of the system, we can say that the associated human activities would provide an input to the business process and there would be scenarios where the business process would be able to continue only after manual decisions. For example, in a banking transaction business process, the customer has to provide a one-time password to continue the process.

A **Human Interaction Management System (HIMS)** should have the capability to handle the life cycle of a human activity, which includes notifying the users who are associated with the activity, collecting information or results from the users, and keeping track of the history of the tasks. The information collected from human interactions is used in process execution and decision making. jBPM has an in-built human task service and can be integrated with any other HIMS. The in-built human task service of jBPM is in compliance with the WS-Human Task specification.

jBPM also provides a form modeling feature that helps business analysts to design user interface forms that can be attached to the jBPM user task, using which information about the completion of the user task (if any) can be collected.

Business activity monitoring

Business Activity Monitoring (BAM) provides online monitoring capabilities for business activities and enables businesses to arrive at key performance pointers by slicing and dicing the activity event data. BAM refers to a general software system that can monitor business processes, but attaching BAM with a BPM-based software system is relatively easy and powerful. In general, BAM software has the capability to show business data using dashboards, enabling business users to create customer reports and charts for performance indicators and trend analysis.

The jBPM core engine stores the process and task history and provides APIs to perform the BAM operations. Further, the jBPM tooling includes the dashboard builder, which enables its users to create custom dashboards from the business process history.

Consider a banking transaction process where a customer carries out a transaction from his/her account to another account. By analyzing the process history logs, the business user can create trend analysis reports of the peak hours of the day at which the maximum number of transactions took place. This analysis can be used to optimize related IT costs.

Another more frequent use case where BAM is applied is the optimization of a business process by avoiding process bottlenecks. Process bottlenecks can be caused by various problems including resource shortage (insufficient supply of staff) or a system not performing up-to-the-mark. A resource shortage problem would be quite obvious, but a system inefficiency may not be very apparent. For example, suppose that the **One-Time Password** (**OTP**) activity in a banking transaction process is often delayed. Let us say it takes more than one minute, which prompts users to perform the transaction again (or use the resend option) and leads to customer dissatisfaction. This also leads to wastage of resources by sending multiple OTPs; in other words, a delayed OTP confuses the customer and the bank further loses its credibility. Such issues are often noticed by banks when diligent customers notify them about the issues. BAMs can make the corrections of these issues more proactive, leading to increased customer satisfaction.

Business process simulation

Business process simulation provides the business user with the ability to see how the business process model created by them would work during runtime. Simulation is a term that goes hand-in-hand with modeling. A model designer would really love to have a simulator to test his model and understand the runtime impacts of his/her model. Business process simulation is the capability to analyze the runtime behavior of business processes. It helps in the optimization of business process models for quality, performance, and resource utilization.

Comparing business process simulation with the more regular models of testing these processes, we find that the user has to manually test these processes after deploying them in a **Quality Assurance** (**QA**) environment. Further, for automated verification, the user would have to rely on a programmer to script the automated test cases. In both cases, the process of verification is not carried out locally with the modeling process and delayed results mean iterations for corrections, adding to the cost of process improvements.

Business process simulation provides business users with the ability to see how the business process model created by them would work during runtime. The users can provide the simulation information, which includes the input to the processes, interrupts, and resource information to be used in the simulation mode. While executing the simulation, the software collects the statistical information about the running process, and this information can be analyzed by the users to optimize their business model.

jBPM is an early adopter and implementer of the **Business Process Simulation Interchange Standard (BPSim)**. BPSim provides a standardized specification that allows business process models to be analyzed using a computer interpretable representation or a meta model. Using jBPM, the end user can provide the metadata for simulation by using the process designer and execute simulations to view the execution paths and the performance metrics associated with the various execution paths.

The business process model and notation

Business Process Model and Notation (BPMN) is the widely accepted standard for business process modeling and provides a graphical notation for specifying business processes in a **Business Process Diagram (BPD)**. It is based on a flowcharting technique very similar to the activity diagrams of **Unified Modeling Language (UML)**. BPMN is maintained by **Object Management Group (OMG)**, and the current version is 2.0 (released in March 2011).

The primary goal of BPMN is to provide a standard notation readily understandable by business stakeholders. These include business analysts who create and refine the processes, technical developers responsible for implementing these processes, and operation managers who monitor and manage the processes. Consequently, BPMN serves as a common language, bridging the communication gap that frequently occurs between business process design and implementation. BPMN also serves as a communication medium between organizations who partner for achieving common business goals, to share functional processes and procedures.

One of the main differences between BPMN and other process definition standards such as **Business Process Execution Language (BPEL)** is that BPMN supports human interaction. Human interaction support provides completeness to business process modeling, as humans are the primary actors in any business organization. Being a specification of visual programming notations, BPMN places considerable emphasis on diagrammatic representations of the elements of the business process model. So, a reader of a BPMN diagram can easily recognize the basic type of elements and understand the business process. BPMN conformance ensures a common visual representation, although it allows variations without dramatically changing the basic look and feel of the diagram.

The detailed explanation of the BPMN 2.0 specification can be found in the specification document.

[BPMN specification documents can be found at http://www.bpmn.org/.]

Conformance to standard BPMN specification defines four types of conformance:

- **Process modeling conformance**: Tools claiming conformance must support BPMN core elements, human interactions, pools, and message flows.

- **Process execution conformance**: Tools claiming conformance must support and interpret operational semantics and activity life cycle as stated in the specification.

- **BPEL process execution conformance**: A special type of process execution conformance that supports BPMN mapping to WS-BPEL.

- **Choreography modeling conformance**: Tools claiming conformance must support choreography diagrams and their elements. Choreography diagrams focus on the collaboration of different groups in activities and the message flow between them.

A jBPM implementation partially claims the first two types of conformance, namely process modeling conformance and process execution conformance. Although jBPM supports all core elements in process modeling, it does not support all the elements described in the specification.

Core elements

The chief constituents of a BPMN diagram, BPMN elements, can be broadly classified into five categories:

- **Flow Objects**: These objects define the behavior of a business process

- **Data**: This represents the data associated in the business process

- **Connecting Objects**: These objects are used to connect the flow objects to each other

- **Swimlanes**: This is used to categorize the flow objects

- **Artifacts**: These provide additional information about the process

Flow Objects

Flow objects are the meat of BPMN; they are used to define how the business process would behave. The following are the major types of flow objects:

- **Events**: An event is something that happens in the course of a business process. These events affect the flow of the model and usually have a trigger for and an impact on the business process. Some examples of events are Start, Stop, and Error. A Start event triggers the start of a process instance and is triggered using an explicit trigger, a message, or a timer. Events can be either signaled from a process (thrown) or can be waited upon (catch).

- **Activities**: Activities are actions performed within a business process. They can be either atomic (called Tasks) or compound (Sub-processes). An example of an activity is a user task or a service task. A user task indicates a human interaction and the action has to be taken manually to complete this task. A task can be complete upon being triggered or can wait for completion (a wait state); for example, a service task is triggered and completed, while a human task is triggered and waited upon for a user to complete the action.

- **Gateways**: Gateways are used as controllers for the branching, merging, forking, and joining of execution paths within a business process. An example is the parallel gateway, which can be used to split the execution path into multiple outgoing branches, with all outgoing branches activated simultaneously. Parallel gateways can also be used to merge branches; they wait for the completion of all incoming branches to complete before triggering the outgoing flow.

Data

Business process execution results in the production of data; for example, in the banking transaction process, the transaction details are data provided by the user regarding the transaction. This data would have to be saved or transferred to another activity for further processing. The following elements are the core of data modeling in BPMN.

- **Data Object**: Data objects are data created as part of a business activity. They can be used for informational purposes to indicate that the activity produces such data.

- **Data Input**: Data inputs specify the input needed for an activity for its completion. For example, an OTP sending task in the banking transaction process would need the details of the customer to send the SMS. So, the input of this activity is mapped from the output of the previous activity or from data that is globally associated with the process instance.

- **Data Output**: Data outputs are data resulting from an activity that have to be mapped to a global process variable or to serve as input to a subsequent activity.

- **Data Store**: A data store provides a mechanism for activities to retrieve or update stored information that persists beyond the scope of the process.

Connecting Objects

Flow objects are connected to each other by using connecting objects. The following are the types of connecting objects:

- **Sequence Flows**: A sequence flow is the basic element used to represent a connection. It is used to connect flow objects and defines the execution order of activities.

- **Message Flows**: Message flows are used to represent information flow across organizational boundaries (a group of activities within an organization or a department, or a role of users).

- **Associations**: Associations are used to represent the association of data or an artifact to a flow object.

Swimlanes

Swimlanes are a visual mechanism for organizing and categorizing activities and form the basis of cross-functional charts using BPMN. They represent an organization, a role or, a system. They are basically of the following two types:

- **Pools**: A pool represents the higher-level categorization of activities. For example, an organization can be represented as a pool. A pool consists of multiple lanes. An example of lanes would be the departments within an organization.

- **Lanes**: A lane represents the categorization within the pool. The lane contains flow objects, connecting objects, and artifacts.

Artifacts

Artifacts are graphical representations that provide supporting information about the process or elements within the process. They don't interfere with the process flow; in other words, we can say that they are opaque from the perspective of process execution. The basic types of artifacts stated in the BPMN specification are as follows:

- **Group**: The Group construct can be used to logically group an arbitrary set of flow objects to show that they logically belong together

- **Text Annotation**: Text annotation can used to associate additional documentation with any element in a BPMN diagram

The banking transaction process is illustrated in the following diagram with the core elements of event, activity, data object, lane, and gateway annotated:

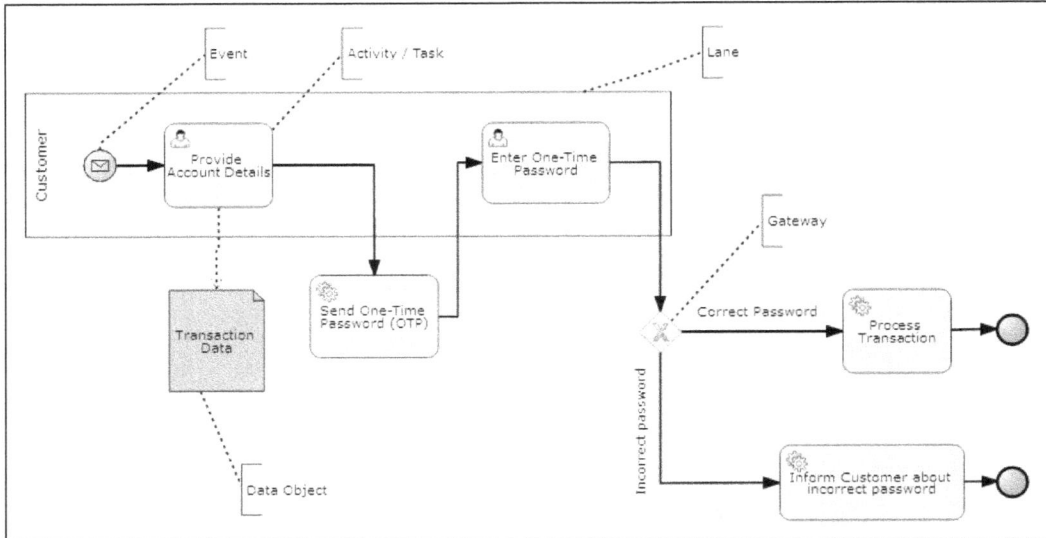

Business process management as applied in industry

BPM is typically used in industries where the following is true:

- The business process is distributed and spans across multiple applications or software systems
- The process involves complex rules that have to be maintained and updated overtime
- There is a need to continuously improve the business process by monitoring the existing activities
- The business is done by the collaboration of multiple stakeholders

With the above considerations, we can apply BPM to any process-oriented domain, such as healthcare, insurance, point of sales, supply chain management, and banking and financial services. A fully functional BPM system gives a huge advantage in terms of the turnaround time for an organization with a new product/service offering or an improved process.

Considering the scope of this book, we can briefly detail some domains in the upcoming sections. A detailed study of these use cases is done in the subsequent chapters, and these use cases are used throughout the book for explaining each and every aspect of jBPM.

Supply chain management

A supply chain is an interconnected set of business procedures and business partners that manages the flow of goods and information from the point of design to the delivery of the product and/or service to the end customer. The supply chain provides a well-coordinated channel to organizations for delivering their products and services to the end customer. Supply chain typical include the following:

- **Suppliers**: They supply the raw materials.
- **Manufacturers**: They manufacture the product.
- **Distributors**: They distributes the product for sale to the retailers.
- **Retailers**: They sell the product to the end customers.
- **Customers**: They buy and use the product. Therefore, a supply chain is a collaboration of multiple functional units, and these groups can be within the same organization or from different business units. Coordinating the jobs and meeting end user expectations in a cost-effective manner is a challenge and is attributed to supply chain management systems. As you can see, the domain is inherently process oriented, with a lot of complex rules and regulations governing each stage of the process, and this is a best-fit domain for applying complete BPM systems.

A few processes to name in the supply chain domain are as follows:

- **Manufacturing flow management**: This deals with the manufacturing of products and/or provision of services
- **Order management**: This meets customer requirements in terms of order fulfillment
- **Customer service management**: This provides real-time information about product availability, shipping dates, and order status

Order management is a core process in the supply chain domain. The process depicted in the next figure is a trivial process but is intended to exemplify the ease and clarity that a business process brings to the table, particularly to a cross-functional business process. From the process diagram itself, any layman can understand the cross-functional units in the organization and, of course, the flow. Thus, the users, both business analysts and operations users, can understand the bigger picture more easily, thereby improving the overall communication and efficiency.

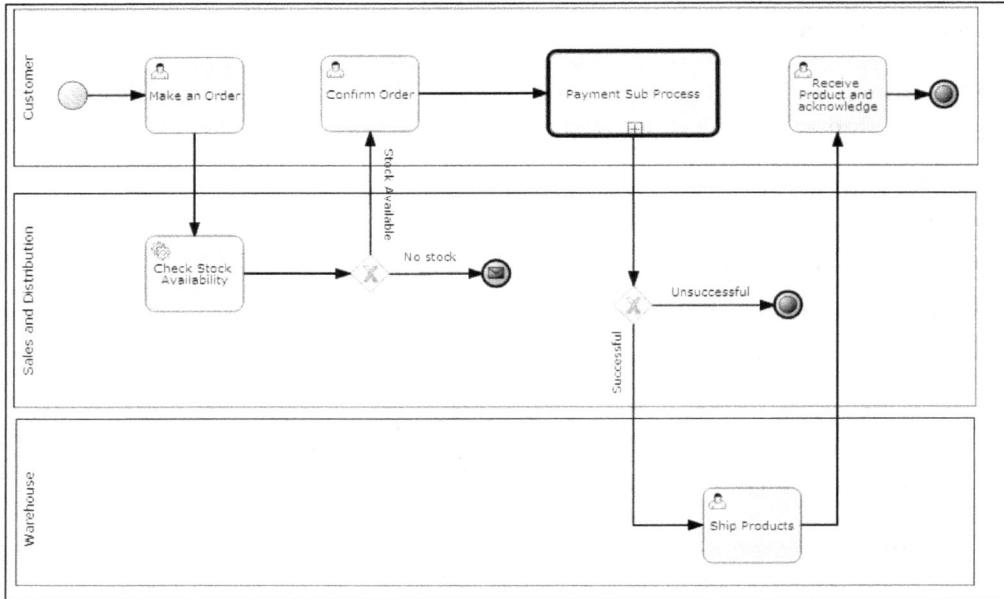

Payment is an activity but has the size and complexity of a separate process. Further, if we apply the rule of responsibility, a payment needs to be clearly abstracted as a separate process that can be reused across the processes of the organization for processing a payment. Thus, we can take a design decision of making payment a sub-process that reduces complexity, enables better handling of exceptional situations, and enables reusability.

Another point I want you to notice in this particular process is the Stop event in **No Stock**; a message is embedded in the termination of the process. This can be used as a signal to another organization process for handling No Stock situations; for example, in this case, **No Stock** would be the signal event to trigger a demand management process.

Banking and financial services

Banking is another domain where BPMs are heavily applied. BPM enables banks to automate their business process such as account opening, loan processing, payments, and transactions. Visibility of processes and compliance to regulations are critical in the banking domain. Banks continuously make innovative changes in their processes and services, and BPMS provides them with the capability to adapt to process improvement initiatives rapidly.

The typical business processes associated with banking are as follows:

- Account opening and maintenance
- Loan processing
- Payments and transaction processing

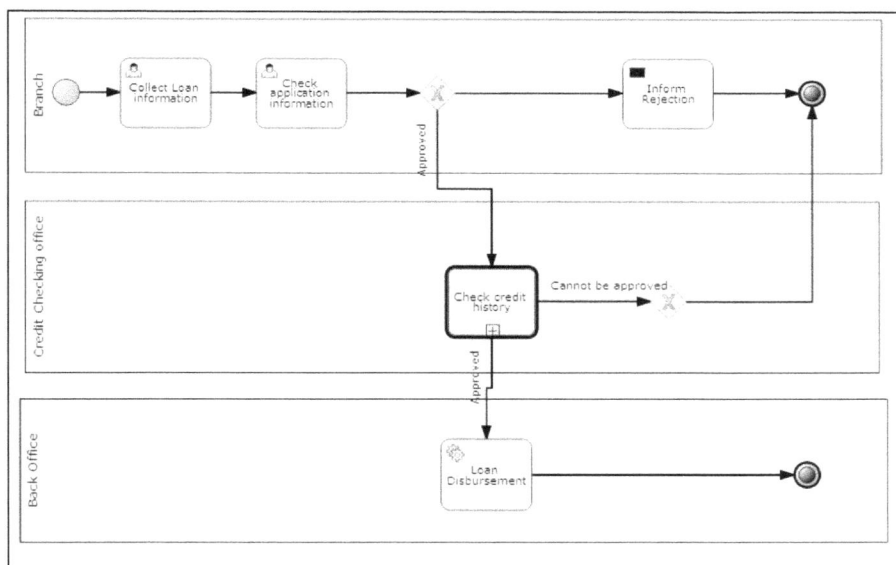

The preceding figure shows a simplified version of loan processing, similar to the supply chain management domain sample process. The clarity it brings in is obvious. However, in the banking domain, the more important aspects are the flexibility, automation, and intelligence options that BPM puts before the business users. The business users can define processes and modify them, thus bringing agility for market promotions; automation helps users to successfully comply with regulatory compliance; and activity monitoring-based analytics help banking organizations to reduce fraud and enhance customer experience.

Design patterns in business process modeling

Design patterns are solutions to commonly occurring problems in their corresponding domain. Business process models try to map general business processes and procedures by using a standard set of elements. The design patterns provided next are solutions for certain commonly occurring problems in business process modeling.

The section focuses on providing an introduction to the design patterns that were identified in the business process modeling domain. Understanding these patterns would help you in identifying solutions easily to the problems you are trying to map using BPMS. The following list of patterns is found in the pattern templates provided by the jBPM designer tools.

Sequence

Sequence is the most basic pattern that occurs in business process modeling, using which tasks to be executed sequentially, one after another, can be modeled.

jBPM supports this by connecting the task using sequence flow connectors, which provide a sequential ordering of the tasks.

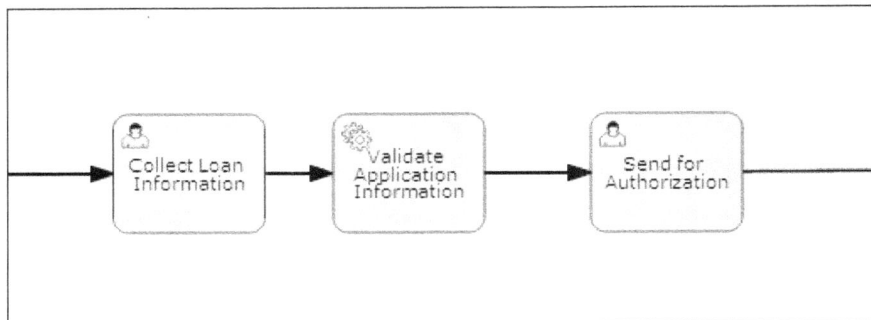

The preceding figure illustrates a part of the loan processing in the banking domain, which represents a sequence pattern problem. Activities such as collecting loan information, validations, and sending for authorization have to be done sequentially; these activities can be allocated to separate human users or even to system automated tasks.

Parallel split

A parallel split pattern is used for splitting the branch of execution to more than two branches in such a way that each of the outgoing branches is executed in parallel.

jBPM supports a parallel split pattern by using a parallel gateway, in which all outgoing branches are activated simultaneously.

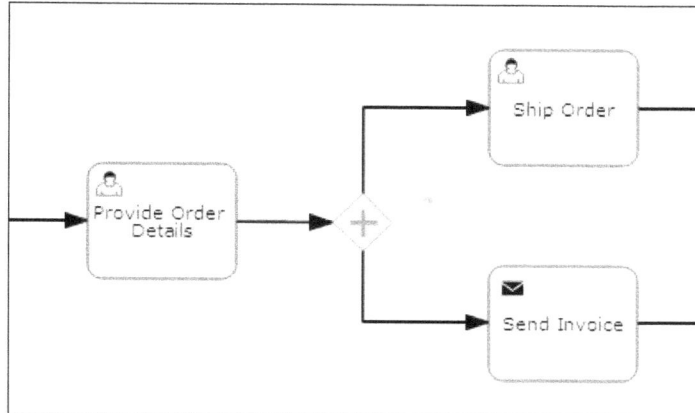

The preceding figure shows a scenario in the order fulfillment process, where a parallel split pattern is applicable. During the ordering process, after providing the order details the shipment of the order and the invoice sending, payments may be parallel paths of processes. The parallel split pattern fits in here, solving the problem.

Synchronization

A synchronization pattern is just the other end of the parallel split design patterns; it merges two or more branches in such a way that the merged branch would run only after the execution of all incoming branches that are to join.

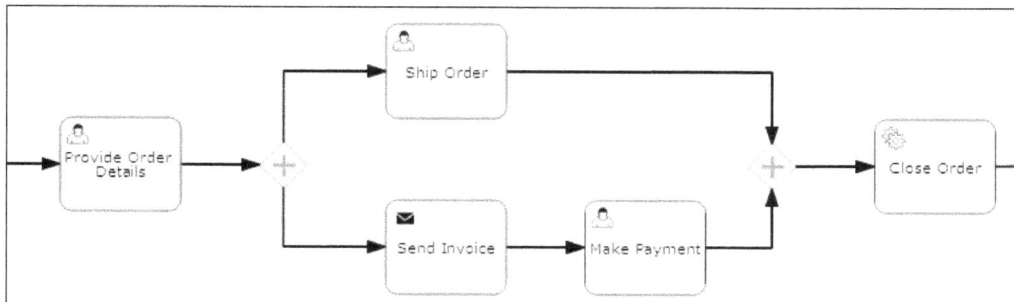

The preceding figure shows the continuation of the example from the parallel split pattern, after the parallel process paths for shipment and payment making are completed, there is a close order activity to be done. Notice that the close order activity has to be done after the completion of both the payment and the shipment activities.

Simple merge

The simple merge pattern provides a way to merge two or more branches in a process definition into a single path of execution. This is particularly useful in scenarios where there are two or more paths to reach a common set of activities. We can avoid duplicating these common set of activities by using a simple merge pattern.

In jBPM, single merge can be realized by using an XOR or exclusive gateway, which awaits one incoming branch to complete before triggering the outgoing flow.

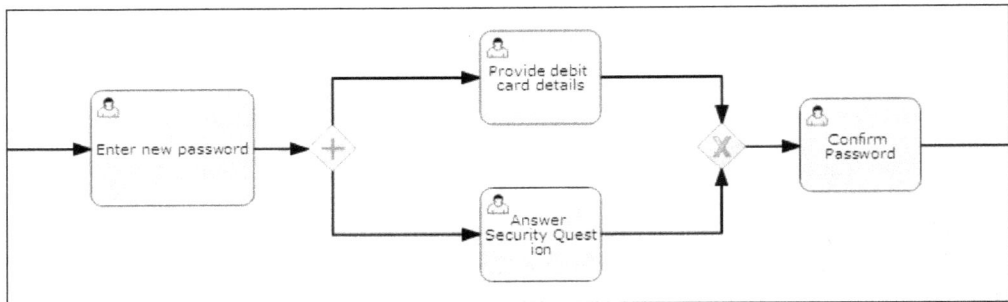

The preceding example process, which deals with changing the password of an online banking customer, fits nicely with the problem statement of a simple merge. As illustrated in the diagram, for ensuring security, the user is given two options, either provide the debit card details or answer a security question. After giving either of these options, password confirmation occurs. The exclusive gateway (before the **Confirm Password** activity) provides control to the confirm password activity after the completion of either of the parallel activities.

Exclusive choice

Exclusive choice patterns provide a solution to diverging a branch into more than one branch (or path of execution) such that after the completion of the incoming branch, the flow of execution is handed over to precisely one branch on the basis of the condition of branching.

In jBPM, this pattern can be implemented using the data-based exclusive (XOR) gateway. The XOR gateway splits and routes the execution to exactly one branch on the basis of the branch condition provided.

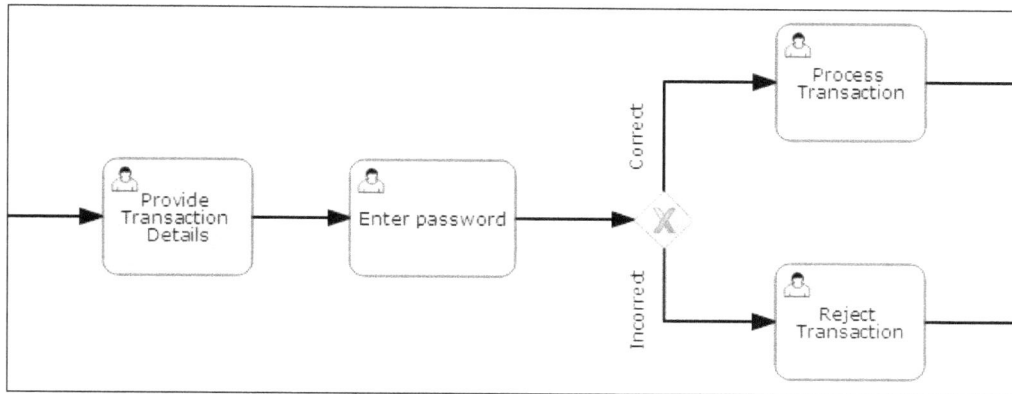

The preceding figure illustrates a scenario for exclusive choice. The sample is a part of a process for making an online transaction. After entering the password based on a validation in the exclusive gateway with the condition, the selection of the path to continue is made. This pattern is similar to a decision box in a flowchart.

Implicit termination

The implicit termination pattern enables a user to terminate the process from any branch. The process engine verifies the completed workitems and decides the termination of the process. This largely avoids clutter because otherwise we have to design the process in such a way that these paths join at a single point of termination. The complexity of such a design would increase with the increasing number of paths in the process.

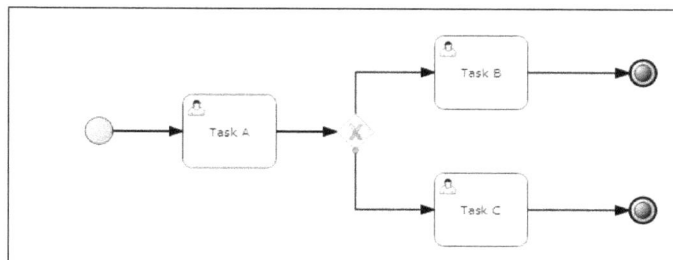

The previous figure shows a process with two terminations or termination events attached. Here, either after completion of **Task B** or after completion of **Task C**, the process terminates.

If implicit termination is not supported by the BPMN implementation, users can achieve the termination by merging the paths to a common termination, and is called as explicit termination.. The following figure shows the explicit termination:

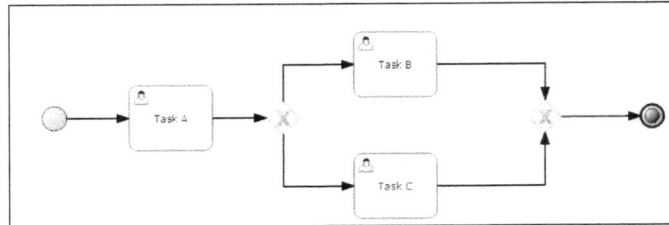

Referring the preceding figure, we can see that a gateway is used to achieve a common termination point. The results of the process execution in the previous two cases are the same, but the second solution (explicit termination) is more complex.

Deferred choice

Deferred choice gives a business process the ability to choose a path on the basis of an interaction with the operating environment. The execution control waits in the decision gateway; the path where the first task is initiated is chosen as the path of execution.

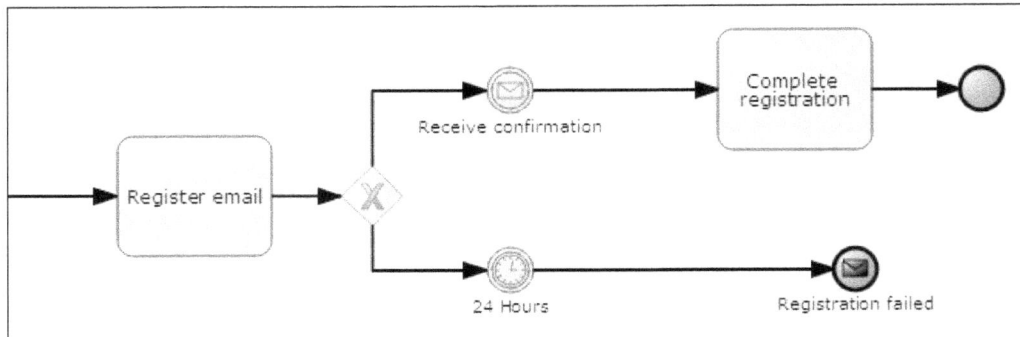

The preceding figure shows the implementation of a deferred choice pattern in an online banking scenario. The process is for enabling the customer to register his/ her e-mail with his/her account to receive updates, account statements, and so on. After the registration, an e-mail is sent to the customer at the registered customer ID to complete registration. If the customer doesn't respond to it within a specified time period (here, **24 Hours**), the registration fails.

Multiple instance without synchronization

By using the **Multiple Instance** (**MI**) facility, we can create multiple instances of a task. These instances are independent of each other. There is no requirement to synchronize the execution flow after the multiple instance execution, unlike a merge.

jBPM allows us to model a multiple instance sub-process, which can be used to implement the MI patterns. An example is shown in the following figure:

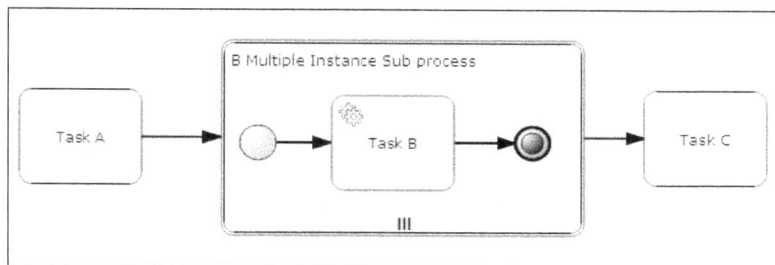

After the execution of **Task A**, the **Task B** execution is done by using a collection expression used to define **B Multiple Instance Sub process**. Multiple instances of **Task B** are executed depending on the number of items provided in the collection expression. **Task C** is executed without waiting for the execution completion of **Task B** (or instances of **Task B**). This pattern is particularly useful when multiple tasks need to be run in a *fire-and forget*-manner.

For example, in a process, we have to send e-mails to a set of users (say subscribers for an incident). Here, the collection expression would be the list of subscribers. The multiple instance task (send e-mail task) would send e-mails to each subscriber.

Synchronized merge

Synchronized merge provides a controlled way for merging a branched execution flow. The execution flow is merged when all incoming "active" branches are completed.

jBPM implements this pattern by using the inclusive gateway. Inclusive gateways, upon splitting, activate one or more branches on the basis of the branching conditions. Upon merging, it waits for all active incoming branches to complete.

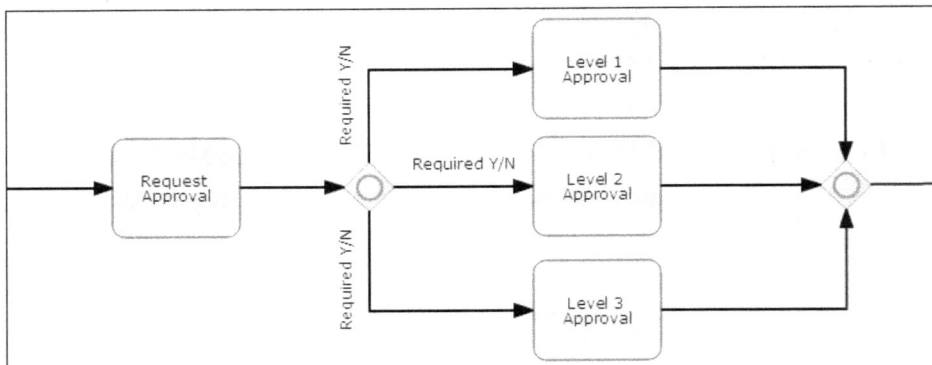

The preceding process illustrates a synchronized merge scenario. Based on the condition for the levels of approval, one or more levels of approval may be required. The second inclusive gateway ensures that the control to the next activity is done after all active approvals are done.

Arbitrary cycle

Arbitrary cycle patterns address the need for repetition of tasks in a process model in an unstructured manner, without the need of explicit constructs such as loop operators. This pattern helps in representing process models that require a cycle in a visually readable format.

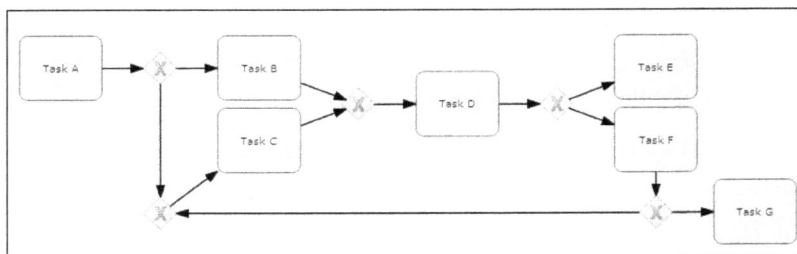

The preceding figure shows that tasks can be cyclically connected using connectors and gateways.

Introducing jBPM

jBPM is an open source BPM suite with a complete tool stack supporting everything right from the modeling and execution to the management and maintenance of business processes. It was released under Apache License 2.0, and was developed and is actively supported by the JBoss community.

The tool stack focuses on serving the following two types of users:

- Business users who model the application and use the application
- Technologists who assist the business users to make the models executable and the application completely functional

The following diagram will give you an overall view of the jBPM tool stack and the set of functionalities that it provides. We will discuss in brief the functional components. Detailed discussions of each of these components will be presented in subsequent chapters.

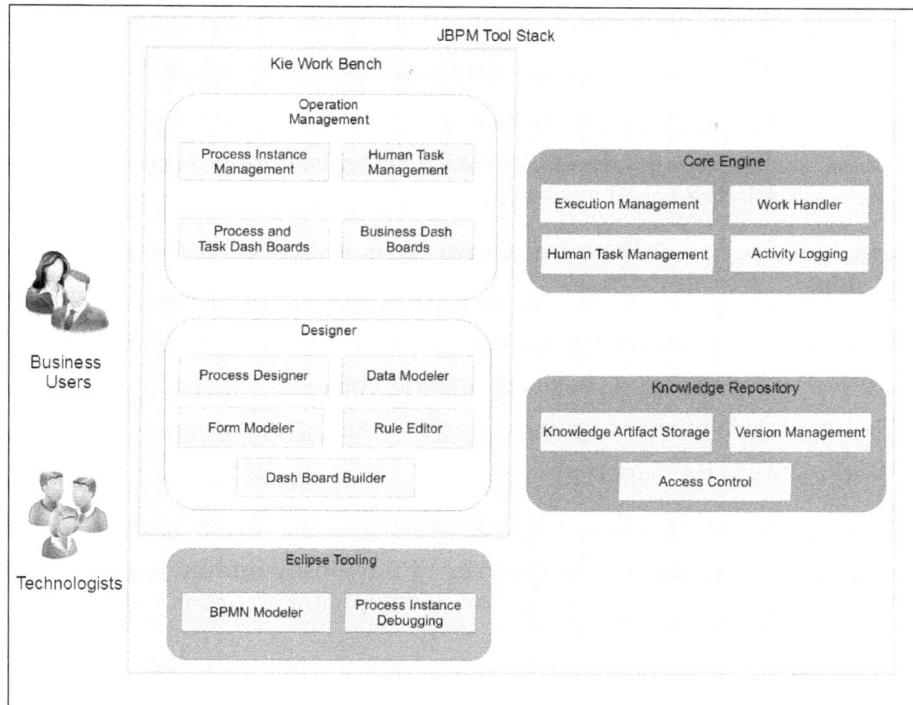

The KIE workbench

Kie is the abbreviated form of "Knowledge is Everything," and the Kie workbench is a combination of every tool in the Drools world for capturing and managing knowledge. It combines capabilities such as authoring of projects, data models, rules, decision tables, test services, process authoring, process runtime execution, process simulation, and human task management.

The Kie workbench provides a web frontend and a complete integrated environment for doing all BPM-related activities. It makes it easier for the user to manage the large set of tools. The underlying architecture is highly modular, and you can integrate each part of this functionality independently to your web application, provided it is UberFire-compliant.

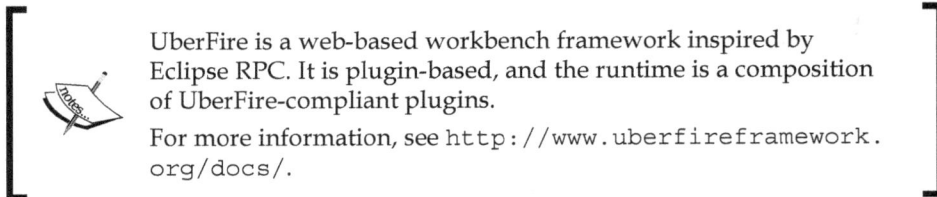

> UberFire is a web-based workbench framework inspired by Eclipse RPC. It is plugin-based, and the runtime is a composition of UberFire-compliant plugins.
>
> For more information, see http://www.uberfireframework.org/docs/.

The importance of Kie workbench is that it also provides a philosophy, a guideline, and a process, in developing knowledge-based systems. It guides you through developing a knowledge-based system in a structured manner.

The knowledge life cycle, as the community calls it, is a cycle consisting of the following steps:

1. **Discover**: The business knowledge required to drive your company.
2. **Author**: Formalize your business knowledge.
3. **Deploy**: Learn how to configure your environment.
4. **Work**: Reduce the paperwork.
5. **Improve**: Enhance your business performance.

The knowledge life cycle shows the maturity of the JBoss community in providing ready-to-use, knowledge-based application development environments.

Process Designer

Process Designer is a web-based rich user interface that allows us to model BPMN2-compliant business processes. Its aim is to provide an intuitive workbench where both business and technical users can model and simulate executable business processes.

Operation management

The output of modeling is executable business processes. Management tooling enables users to manage the execution of processes, monitor the execution, and report it.

jBPM provides web-based tooling for process management, monitoring, and reporting. Web-based tooling in fact uses the core engine APIs. These APIs are exposed via **Representational State Transfer** (**REST**), **Java Message Service** (**JMS**), and **Contexts and Dependency Injection** (**CDI**) interfaces so that they can be integrated with other enterprise software systems.

Eclipse tooling

Although as a BPM suite, jBPM focuses on business users with a large set of web-based tooling for business users, the community firmly keeps their feet on the ground, with explicit tooling for technical users who are relied upon to build complex business functionalities.

Eclipse-based tooling is a set of plugins to the Eclipse IDE and allows technical users to integrate the jBPM environment to the software development environment. The Eclipse tooling provides features such as the following:

- Wizard for creating a new jBPM project
- Graphical modeler for BPMN 2.0 processes
- Runtime support (for selecting the jBPM runtime to use)
- Debugging, the current states of the running processes can be inspected and visualized during execution
- Process instance view, providing information showing all running process instances and their state
- Audit view showing the audit log
- Synchronizing with the Kie workbench repositories, enabling collaboration between eclipse tooling users and the web-based users

Core engine

The core of jBPM is a business process execution engine built in Java. It is lightweight and is easily embeddable in any Java application as a dependency (as dependent libraries). The core engine is designed as a standard-based, pluggable, and highly extensible component. It supports the native BPMN 2.0 specification.

The knowledge repository

The knowledge repository is where we store all the process definitions and related artifacts. BPM is a continuous process; business processes continuously evolve, and it is important to keep track of the updated processes and provide multiple versions of the processes. jBPM uses Drools Guvnor as the knowledge repository and provides the following:

- Persistence storage for processes and related artifacts such as processes, rules, data models, and forms

- Deploying selected processes

- Authentication and authorization

- Categorization and searching of knowledge artifacts

- Scenario testing to make sure you don't break anything when we update processes

- Features for collaborative development of business processes, such as comments and change notifications

The business logic integration platform

It is also important to understand that jBPM is a part of a package provided by JBoss, the **Business Logic Integration Platform (BLIP)** that consists of the following:

- Drools Guvnor (Business rules manager)

- Drools Expert (Rule engine)

- jBPM (Process Management)

- Drools Fusion (Event processing/Temporal reasoning)

Together they form a complete solution for knowledge-based application development and management for enterprise solutions.

BILP can be visualized as a rapid enterprise application development platform from which applications can be solely built by modeling rules, business process flows, events, data models, and forms with very little help from the technical users. The technology and its tools have dual focus: they explicitly serve business users, enable these users to express requirements directly by modeling, and help to engage them in the application development process, guided by technical users who make the models fully functional and realize the software application with all its qualities (nonfunctional requirements).

BLIP integrates the following three paradigms in business language-driven applications:

- **Business Rule**: This represents knowledge in the business domain
- **Business Processes**: This represents activities performed in the business domain
- **Event Stream Processing**: This adds temporal reasoning as part of the knowledge

These paradigms, although developed as three different streams, as obvious from the definitions itself, are interrelated in a business solution context. No wonder in most of the solution architectures, these three form the cornerstones of business modeling.

These paradigms are integrated into a unified platform, where all features of one module are leveraged by the other modules. The decision-making (from the knowledge base) capabilities in Drools Expert are used by Drools Fusion; the event processing capabilities of Drools Fusion can be used by the BPM suite jBPM; and jBPM uses Drools Expert internally for executing business rule tasks.

The following figure depicts this integration and the interactions between the modules in the platform. Drools Guvnor is used for designing processes, events, rules, and related artifacts; storing them; and providing integrated testing facilities. As an integrated platform, BLIP allows interactions between rules, processes, and events.

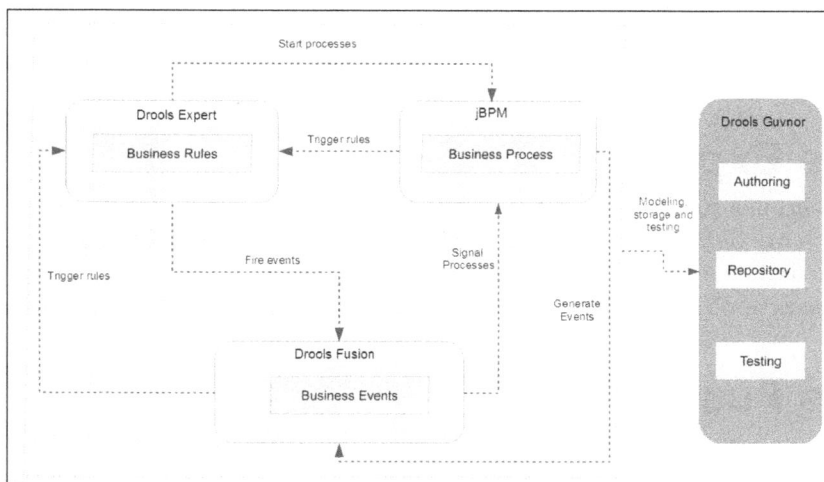

As depicted in the preceding diagram:

- **Drools Expert** can start a business process or create events based on its rule inference
- **Drools Fusion** can trigger rules or signal processes based on its temporal reasoning capacity based on the incoming pattern of events; and jBPM can trigger rules as one of its activities or generate events based on the business activity

We have already discussed the overview of jBPM; now, let us briefly discuss the other component in the platform in subsequent sections.

Drools Guvnor

We have already discussed the knowledge repository in the core concepts section of jBPM, and you guessed correct, jBPM uses Drools Guvnor as its knowledge repository.

Drools Guvnor provides the following features:

- Assistive authoring for knowledge artifacts such as process definition, rule definition, events, and related artifacts
- Access control and security for using these artifacts
- Version management
- Assisted deployment and distribution of the artifacts to the runtime environment
- Integrated testing

Drools Expert

Drools Expert is a rule engine. The knowledge is stored in the knowledge base as rules, which are defined in a declarative language, **Drools Rule Language** (**DRL**). The engine matches the incoming data or facts against these rules to reach a decision and execute actions attached to the inferences/decisions.

Drools Fusion

Drools Fusion is an event processing engine used to detect and select the events of interest (business interest) from multiple streams or an event cloud. Fusion works in tight integration with Drools Expert.

Drools Fusion provides the following features:

- Temporal reasoning, which allows the definition and inference of business rules based on time factor of events
- Scheduling and delaying of actions to be taken on inferences

Working together

Now that we have discussed the tools, we can now move on to visualize how these tools work together to make a business solution that is completely rule-driven. The following figure provides a bird's eye view of a possible solution architecture where we use these tools together.

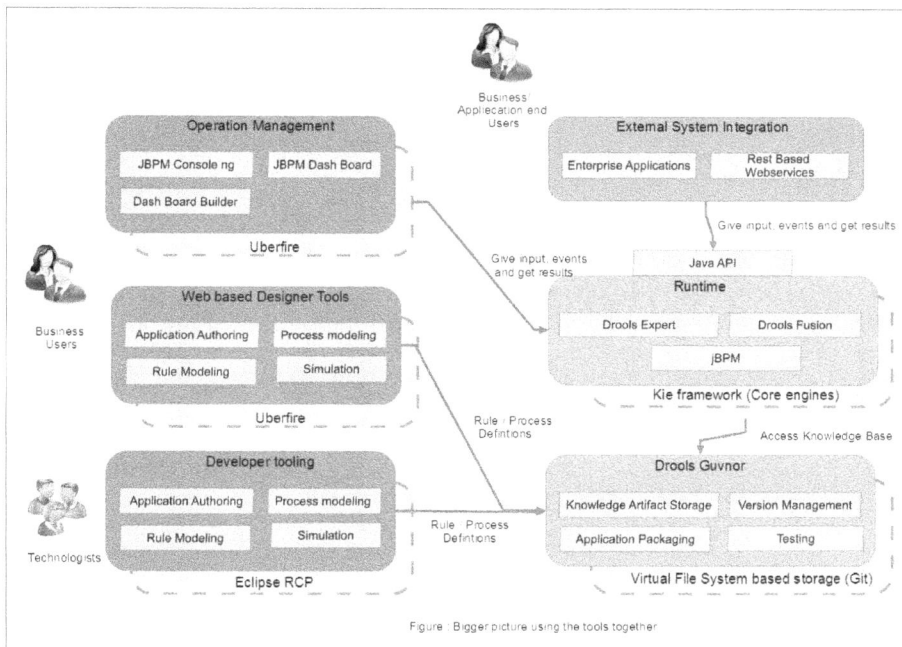

Figure : Bigger picture using the tools together

Together, they enable business analysts to write and manage applications by mapping business scenarios to process definitions, rule definitions for decisions, and rules including temporal logic.

1. The designer tools enable business analysts to map the business scenarios required for a solution to rules, processes, events, or related artifacts.

2. The applications/users can interact with the solution by using the operation management tooling provided by jBPM, or they can use the RESTful web service interface to access the same from their proprietary applications.

3. The runtime integrates the business processing, rule inference, and temporal reasoning capabilities to infer decisions and take actions for a business scenario.

4. The knowledge repository module provides storage, version management, deployment, and testing of the applications developed.

Another important thing I want to highlight from the preceding figure is the technology platforms used by the individual tools. The web-based user interface for process designing and operation management is developed using the UberFire framework; the Eclipse-based tooling is based on eclipse RCP; the runtime is based on the Kie framework; and Guvnor is based on virtual file system-based storage and by default uses Git. The details of these technology platforms so as to customize and extend jBPM will be discussed in *Chapter 7, Customizing and Extending jBPM*.

Red Hat also provides a commercially supported flavor of this package and is called Red Hat JBoss BPM Suite.

Summary

The goal of this chapter was to give you a picture of the world of BPM. We explored the basic concepts, the standard, patterns, use cases where BPM is applied in the industry, and its benefits. We also explored the jBPM tool stack and the bigger picture of BLIP. These concepts and tools will be discussed in detail throughout this book.

In the next chapter, we will discuss how to install the jBPM tool stack and create our first business process-centric application using jBPM.

2
Building Your First BPM Application

Let us now build our first BPM application by using the jBPM tool stack. This chapter will guide you through the following topics:

- Installing the jBPM tool stack
- Hacking the default installation configurations
- Modeling and deploying a jBPM project
- Embedding jBPM inside a standalone Java project

A step towards the goal of this book, the chapter gives you the hands-on flexibility of the jBPM tool stack and provides information on hacking the configuration and playing around.

Installing the jBPM tool stack

A jBPM release comes with an installation zip file, which contains the essentials for the jBPM environment and tools for building a demo runtime for easy hands-on management of the jBPM runtime environment.

For downloading jBPM:

1. Go to `http://jboss.org/jbpm` | **Download** | **Download jBPM 6.2.0.Final** | **jbpm-6.2.0.Final-installer-full.zip**.

   Use the latest stable version. The content of the book follows the 6.2.0 release.

2. Unzip and extract the installer content and you will find an `install.html` file that contains the helper documentation for installing a demo jBPM runtime with inbuilt projects.

> jBPM installation needs JDK 1.6+ to be installed and set as `JAVA_ HOME` and the tooling for installation is done using ANT scripts (ANT version 1.7+).

The tooling for installation is basically an ANT script, which is a straightforward method for installation and can be customized easily. To operate the tooling, the ANT script consists of the ANT targets that act as the commands for the tooling. The following figure will make it easy for you to understand the relevant ANT targets available in the script. Each box represents an ANT target and helps you to manage the environment. The basic targets available are for installing, starting, stopping, and cleaning the environment.

To run the ANT target, install ANT 1.7+, navigate to the installer folder (by using the shell or the command line tool available in your OS), and run the target by using the following command:

```
ant <targetname>
```

> **Downloading the example code**
>
> You can download the example code files from your account at `http://www.packtpub.com` for all the Packt Publishing books that you have purchased. If you purchased this book elsewhere, you can visit `http://www.packtpub.com/support` and register to have the files e-mailed directly to you.

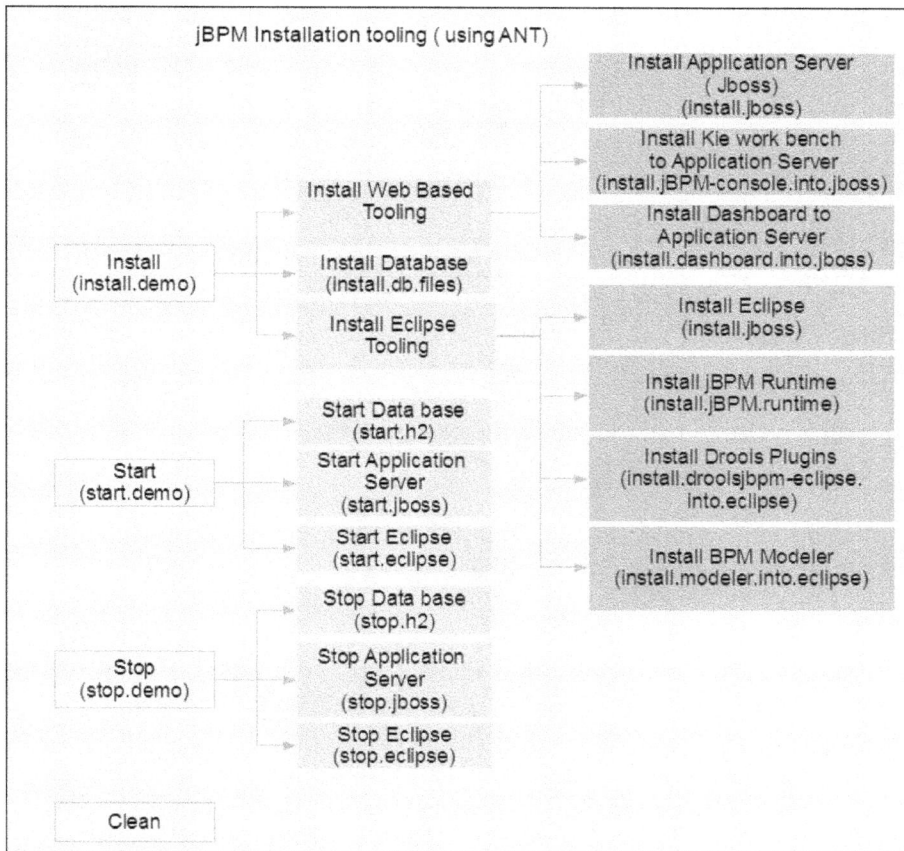

The jBPM installer comes with a default demo environment, which uses a basic H2 database as its persistence storage. The persistence of jBPM is done using Hibernate; this makes it possible for jBPM to support an array of popular databases including the databases in the following list:

Hibernate or Hibernate ORM is an object relational mapping framework and is used by jBPM to persist data to relation databases. For more details, see http://hibernate.org/.

Databases Supported	Details
DB2	http://www-01.ibm.com/software/in/data/db2/
Apache Derby	https://db.apache.org/derby/
H2	http://www.h2database.com/html/main.html
HSQL Database Engine	http://hsqldb.org/
MySQL	https://www.mysql.com/
Oracle	https://www.oracle.com/database/
PostgreSQL	http://www.postgresql.org/
Microsoft SQL Server Database	http://www.microsoft.com/en-in/server-cloud/products/sql-server/

For installing the demo, use the following command:

```
ant install.demo
```

The install command would `install` the web tooling and the Eclipse tooling, required for modeling and operating jBPM.

```
ant start.demo
```

This command will start the application server (JBoss) with the web tooling (the Kie workbench and dashboard) deployed in it and the eclipse tooling with all the plugins installed.

> Refer `install.html`, the installation documentation included in the installer archive for common errors that can occur during installation; it guides you to the solutions. The installation document is quite elaborate; please refer to it for a more detailed understanding of the installation process.

That's it for the installation! Now, the JBoss application server should be running with the Kie workbench and dashboard builder deployed.

You can now access the Kie workbench demo environment by using the URL and log in by using the demo admin user called `admin` and the password `admin`:

```
http://localhost:8080/jbpm-console.
```

Customizing the installation

The demo installation is a sandbox environment, which allows for an easy installation and reduces time between you getting the release and being able to play around by using the stack. Even though it is very necessary, when you get the initial stuff done and get serious about jBPM, you may want to install a jBPM environment, which will be closer to a production environment. We can customize the installer for this purpose. The following sections will guide you through the options available for customization.

Changing the database vendor

The jBPM demo sandbox environment uses an embedded H2 database as the persistence storage. jBPM provides out of the box support for more widely used databases such as MySQL, PostgreSQL, and so on. Follow these steps to achieve a jBPM installation with these databases:

1. Update the `build.properties` file available in the root folder of the installation to choose the required database instead of H2. By default, configurations for MySQL and PostgreSQL are available. For the support of other databases, check the hibernate documentation before configuring.

2. Update `db/jbpm-persistence-JPA2.xml`, and update the `hibernate.dialect` property with an appropriate Hibernate dialect for our database vendor.

3. Install the corresponding JDBC driver in the application server where we intend to deploy jBPM web tooling.

Manually installing the database schema

By default, the database schema is created automatically by using the Hibernate autogeneration capabilities. However, if we want to manually install the database schemas, the corresponding DDL scripts are available in `db\ddl-scripts` for all major database vendors.

Creating your first jBPM project

jBPM provides a very structured way of creating a project. The structure considers application creation and maintenance for large organizations with multiple departments. This structure is recommended for use as it is a clean and secure way of manning the business process artifacts. The following image details the organization of a project in jBPM web tooling (or the Kie workbench).

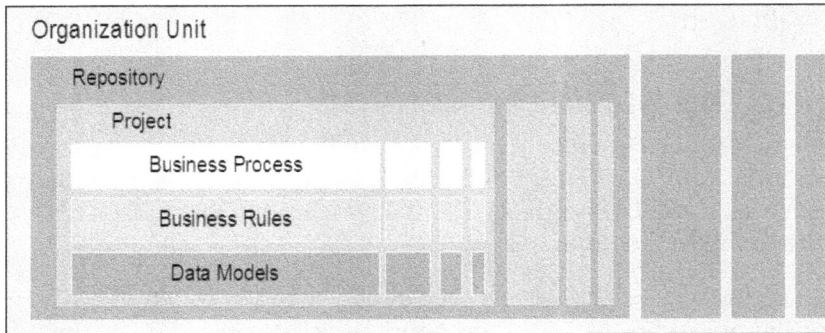

The jBPM workbench comes with an assumption of one business process management suite for an organization. An organization can have multiple organization units, which will internally contain multiple projects and form the root of the project, and as the name implies, it represents a fraction of an organization. This categorization can be visualized in any business organization and is sometimes referred as departments. In an ideal categorization, these organization units will be functionally different and thus, will contain different business processes. Using the workbench, we can create multiple organization units.

The next categorization is the repository. A repository is a storage of business model artifacts such as business processes, business rules, and data models. A repository can be mapped to a functional classification within an organization, and multiple repositories can be set up if these repositories run multiple projects; the handling of these project artifacts have to be kept secluded from each other (for example, for security).

Within a repository, we can create a project, and within a project, we can define and model business process artifacts. This structure and abstraction will be very useful to manage and maintain BPM-based applications.

Let us go through the steps in detail now.

After installation, you need to log into the Kie workbench. Now, as explained previously, we can create a project. Therefore, the first step is to create an organizational unit:

1. Click through the menu bars, and go to **Authoring | Administration | Organizational Units | Manage Organizational Units**.

    This takes you to the Organizational Unit Manager screen; here, we can see a list of organizational units and repositories already present and their associations.

2. Click **Add** to create an organizational unit, and give the name of the organization unit and the user who is in charge of administering the projects in the organization unit.

Add New Organizational Unit

Organizational Unit Information * is required

* Name

Mastering-jBPM

Owner

admin

● Ok   Cancel

3. Now, we can add a repository, navigate through the menus, and go to **Authoring | Administration | Repositories | New Repository**.

4. Now, provide a name for the repository, choose the organization unit, and create the repository.

Create Repository

Repository Infomation * is required

* Repository Name

Master-Repo

* Organizational Unit

Mastering-jBPM

Cancel   Create

Creating the repository results in (internally) creating a Git repository. The default location of the Git repository in the workbench is `$WORKING_DIRECTORY/.niogit` and can be modified by using the following system property: `-Dorg.uberfire.nio.git.dir`.

1. Now, we can create a project for the organization unit. Go to **Authoring | Project Authoring | Project Explorer**. Now, choose your organization unit (here, **Mastering-jBPM**) from the bread crumb of project categorization.

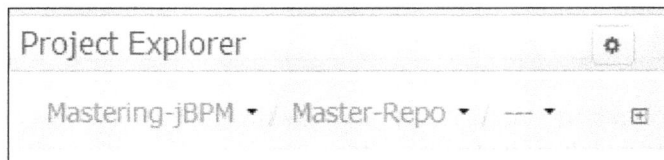

Project Explorer   ⚙

Mastering-jBPM ▼ / Master-Repo ▼ / --- ▼   ⊞

2. Click **New Item** and choose **Project**. Now, we can create a project by entering a relevant project name.

Create new Project

* Resource Name   Introductory-Project

Location   default://master@Master-Repo/

⊕ Ok   Cancel

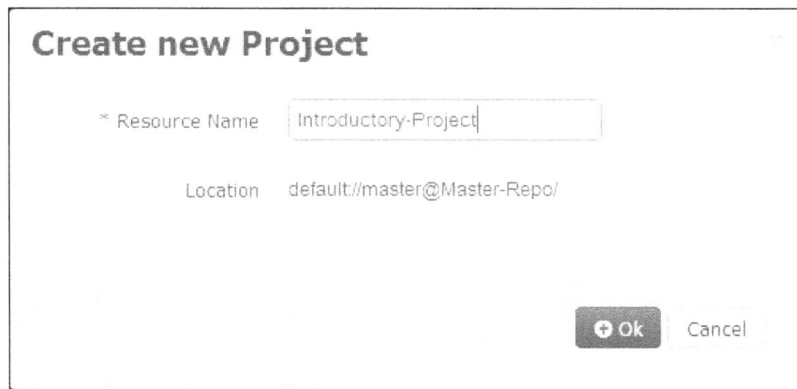

3. This takes you to the new project wizard as shown in the following figure:

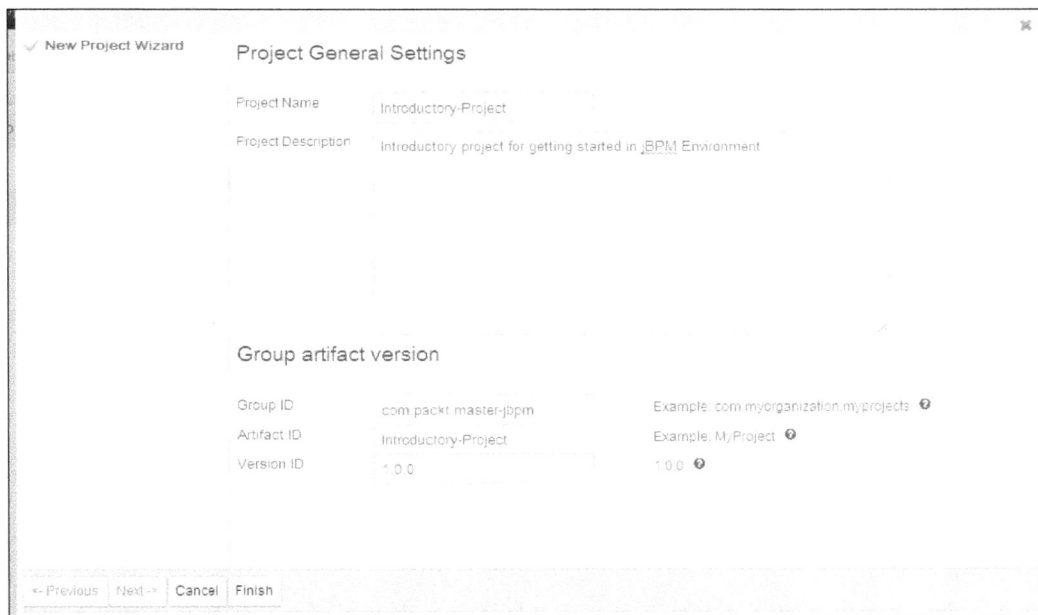

New Project Wizard   Project General Settings

Project Name   Introductory-Project

Project Description   Introductory project for getting started in JBPM Environment

Group artifact version

Group ID   com.packt.master-jbpm   Example: com.myorganization.myprojects ❷

Artifact ID   Introductory-Project   Example: MyProject ❷

Version ID   1.0.0   1.0.0 ❷

← Previous   Next →   Cancel   Finish

4. This gives details like project name and a brief summary of the project, and more importantly, gives the group ID, artifact ID, and version ID for the project. Further, **Finish** the creation of new project.

Those of you who know Maven and its artifact structure, will now have got an insight on how a project is built. Yes! The project created is a Maven module and is deployed as one. We will get into the detail of this is in the coming chapters.

Business Process Modeling

Therefore, we are ready to create our first business process model by using jBPM.

1. Go to **New Item | Business Process**:

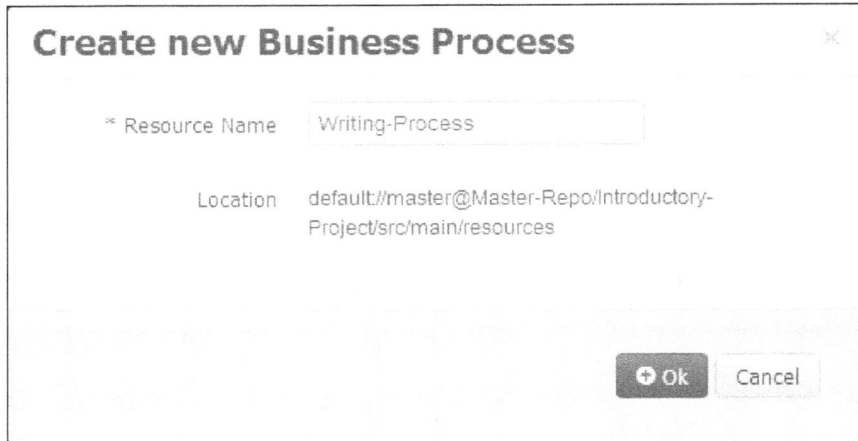

Provide the name for the business process; here, we are trying to create a very primitive process as an example.

2. Now, the workbench will show you the process modeler for modeling the business process. Click the zoom button in the toolbar, if you think you need more real estate for modeling (highlighted in red in the following image):

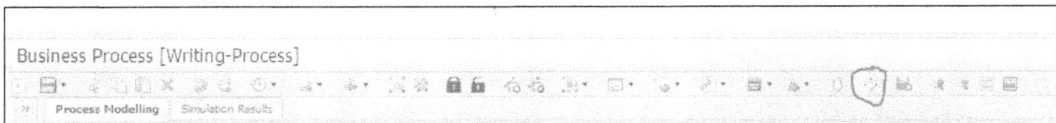

Basically, the workbench can be divided into five parts:

- Toolbar (on the top): It gives you a large set of tools for visual modeling and saving the model.
- Object library (on the left side of the canvas): It gives you all the standard BPMN construct stencils, which you can drag and drop to create a model.

- Workspace (on the center): You get a workspace or canvas on which you can draw the process models. The canvas is very intuitive; if you click on an object, it shows a tool set surrounding it to draw the next one or guide to the next object.

- Properties (on the right side of the canvas): It gives the property values for all the attributes associated with the business process and each of its constructs.

- Problems (on the bottom): It gives you the errors on the business process that you are currently modeling. The validations are done on save, and we have provisions to have autosave options.

The following screenshot shows the process modeler with all the sections described:

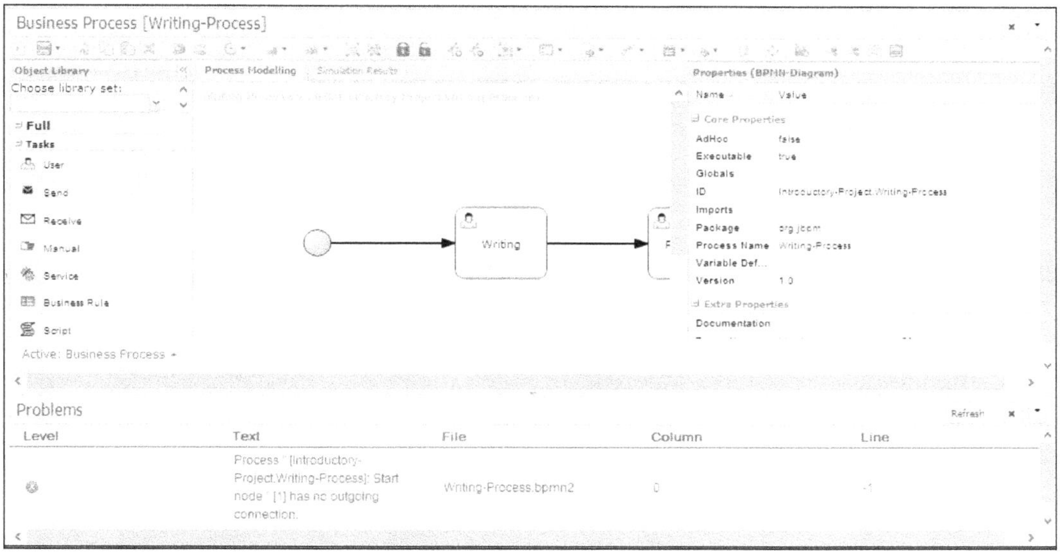

Therefore, we can start modeling out first process. I assume the role of a business analyst who wants to model a simple process of content writing. This is a very simple process with just two tasks, one human task for writing and the other for reviewing.

We can attach the actor associated with the task by going to the **Properties** panel and setting the actor. In this example, I have set it as **admin**, the default user, for the sake of simplicity.

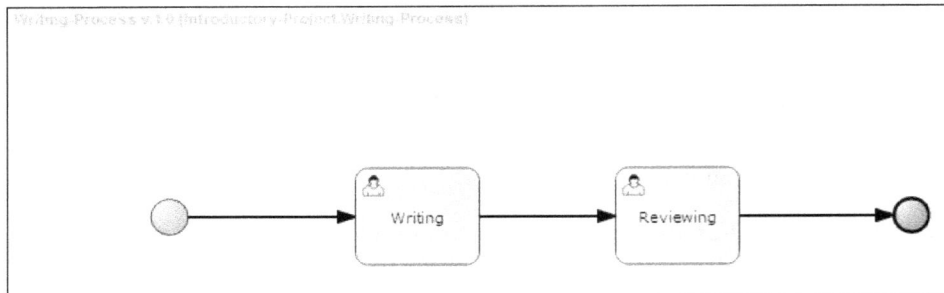

Now, we can save the project by using the **Save** button; it asks for a check-in comment, which provides the comment for this version of the process that we have just saved. Process modeling is a continuous process, and if properly used, the check-in comment can helps us to keep track on the objectives of process updates.

Building and deploying the project

Even though the project created is minuscular with just a sample project, this is fully functional! Yes, we have completed a business process, which will be very limited in functionality, but with its limited set of functionalities (if any), it can be deployed and operated.

Go to **Tools** | **Project Editor**, and click **Build & Deploy**, as shown in the following screenshot:

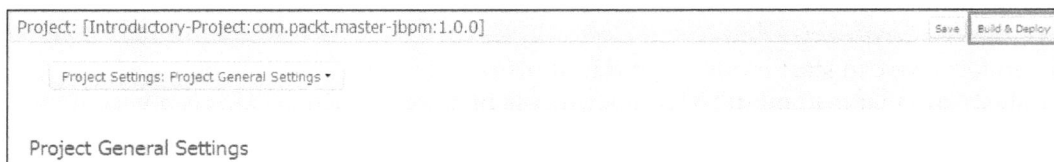

To see the deployment listed, go to **Deploy | Deployments** to see **Deployment Units**, as shown in the following screenshot:

Deployment Units			
Deployment	**Group ID**	**Artifact**	**Version**
com.packt.master-jbpm:Introductory-Project:1.0.0	com.packt.master-jbpm	Introductory-Project	1.0.0

This shows the effectiveness of jBPM as a rapid application builder using a business process. We can create, model, and deploy a project within a span of minutes.

Running your first process

Here, we start the operation management using jBPM. Now, we assume the role of an operational employee. We have deployed a process and have to create a process instance and run it.

1. Go to **Process Management | Process Definitions**. We can see the details of the process definitions deployed in the following screenshot:

Process Definitions			Details	
Name	**Version**	**Actions**	Definition Id	Introductory-Project.Writing-Process
Writing-Process	1.0	⊙ Q	Definition Name	Writing-Process
			Deployment	com.packt.master-jbpm:Introductory-Project:1.0.0
			Human Tasks	
			Human Task Count	0
			User and Groups	admin - Writing admin - Reviewing
			Sub Processes	
			Process Variables	

2. Click **New Instance** and start the process. This will start a process instance.

3. Go to **Process Management | Process Instances** to view the process instance details and perform life cycle actions on process instances.

   The example writing process consists of two human tasks. Upon the start of the process instance, the Write task is assigned to the admin. The assigned task can be managed by going to the task management functionality.

4. Go to **Tasks | Tasks List**:

In **Tasks List**, we can view the details of the human tasks and perform human task life cycle operations such as assigning, delegating, completing, and aborting a task.

Embedding jBPM in a standalone Java application

The core engine of jBPM is a set of lightweight libraries, which can be embedded in any Java standalone application. This gives the enterprise architects the flexibility to include jBPM inside their existing application and leverage the functionalities of BPM.

This section will cover how to programmatically start the runtime engine and start a process instance, and will guide you in writing automated tests for BPMN processes.

Modeling the business process using Eclipse tooling

Upon running the installation script, jBPM installs the web tooling as well as the Eclipse tooling. The Eclipse tooling basically consists of the following:

- jBPM project wizard: It helps you to create a jBPM project easily
- jBPM runtime: An easy way of choosing the jBPM runtime version; this associates a set of libraries for the particular version of jBPM to the project
- BPMN Modeler: It is used to model the BPMN process
- Drools plugin: It gives you the debugging and operation management capabilities within Eclipse

Creating a jBPM project using Eclipse

The Eclipse web tooling is available in the installer root folder. Start Eclipse and create a new jBPM Maven project:

1. Go to **File | New Project | jBPM Project** (Maven).
2. Provide the project name and location details; now, the jBPM project wizard will do the following:
    - Create a default jBPM project for you with the entire initial configuration setup
    - Attach all runtime libraries
    - Create a sample project
    - Set up a unit testing environment for the business process

3. The following screenshot shows the jBPM project wizard.

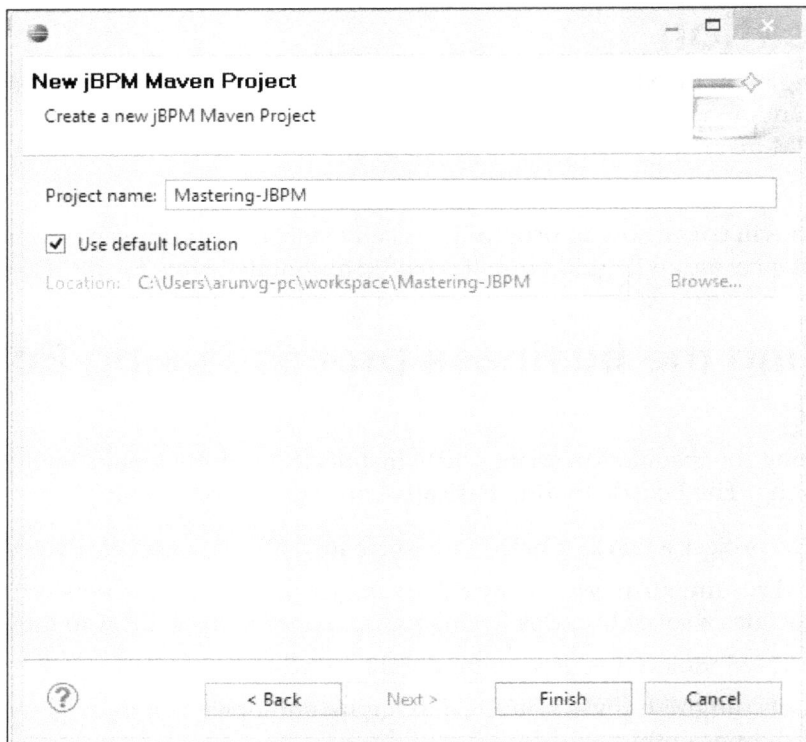

4. The Eclipse workbench is considerably similar to the web tooling workbench; a screenshot is shown as follows:

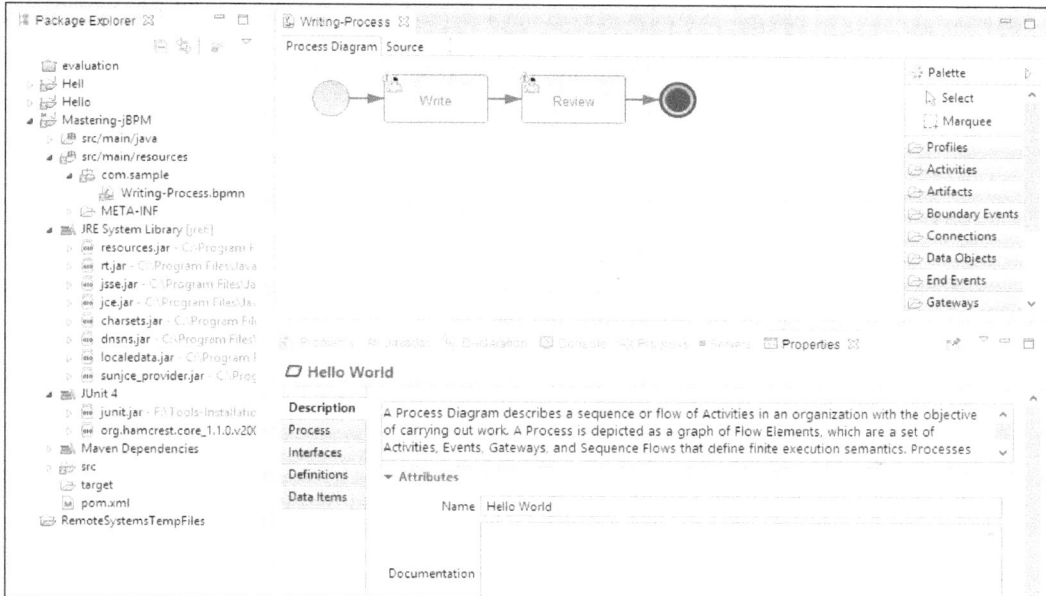

Similar to web tooling, it contains the toolbox, workspace, palette showing the BPMN construct stencils, and the property explorer.

We can create a new BPMN process by going to **New Project Wizard** and selecting **jBPM | BPMN2 Process**. Give the process file name and click **Finish**; this will create a default BPMN2 template file. The BPMN2 modeler helps to visually model the process by dragging and dropping BPMN constructs from the palette and connecting them using the tool set.

Deploying the process programmatically

For deploying and running the business process programmatically, you have to follow these steps:

> KIE is the abbreviation for **Knowledge Is Everything**.

1. **Creating the knowledge base**: Create the Kie Services, which is a hub giving access to the services provided by Kie:

   ```
   KieServices ks = KieServices.Factory.get();
   ```

   Using the Kie Service, create Kie Container, which is the container for the knowledge base:

   ```
   KieContainer kContainer = ks.getKieClasspathContainer();
   ```

   Create and return the knowledge base with the input name:

   ```
   KieBase kbase = kContainer.getKieBase("kbase");
   ```

2. **Creating a runtime manager**: The runtime manger manages the runtime build with knowledge sessions and Task Service to create an executable environment for processes and user tasks.

   Create the JPA entity manager factory used for creating the persistence service, for communicating with the storage layer:

   ```
   EntityManagerFactory emf =
   Persistence.createEntityManagerFactory(

   "org.jbpm.persistence.jpa");
   ```

   Create the runtime builder, which is the ds1 style helper to create the runtime environment:

   ```
   RuntimeEnvironmentBuilder builder =
   RuntimeEnvironmentBuilder.Factory.get()

   .newDefaultBuilder().entityManagerFactory(emf)

   .knowledgeBase(kbase);
   ```

Using the runtime environment, create the runtime manager:

```
RuntimeManager RuntimeManager =
RuntimeManagerFactory.Factory.get()

.newSingletonRuntimeManager(builder.get(),
"com.packt:introductory-sample:1.0");
```

3. **Creating the runtime engine**: Using the runtime manager, creates the runtime engine that is fully initialized and ready for operation:

```
RuntimeEngine engine = manager.getRuntimeEngine(null);
```

4. **Starting the process**: Using the runtime manager, create a knowledge session and start the process:

```
KieSession ksession = engine.getKieSession();
ksession.startProcess("com.sample.bpmn.hello");
```

This creates and starts a process instance.

From the runtime manager, we can also access the human task service and interact with its API.

Go to **Window** | **Show View** | **Others** | **Drools** | **Process Instances** to view the created process instances:

```
Console  Tasks  Process Instances   JUnit
  [1]= RuleFlowProcessInstance (id=5904)
    id= 1
  ▷  processName= "Hello World" (id=5908)
  ▷  processId= "com.sample.bpmn.hello" (id=5907)
  ▷  nodeInstances= Object[] (id=5917)
```

Writing automated test cases

jBPM runtime comes with a test utility, which serves as the unit testing framework for automated test cases. The unit testing framework uses and extends the capabilities of the JUnit testing framework and basically provides the JUnit life cycle methods and the jBPM runtime environment for testing and tearing down the runtime manager after test execution. Helper methods manage the knowledge base and the knowledge session, getting workitem handlers and assertions to assert process instances and various stages.

For creating a JUnit test case, create a class extending `org.jbpm.test.JbpmJUnitBaseTestCase`

We can initialize the jBPM runtime by using the previous steps and assert using the helper methods provided by `org.jbpm.test.JbpmJUnitBaseTestCase`.

For example, we assert the completion of a process as follows:

```
assertProcessInstanceCompleted(processInstance.getId(), ksession);
```

The code for the introductory sample project is attached with the downloads associated with this book.

Change management – updating deployed process definitions

We have modeled a business process and deployed it; the application end users will create process instances and fulfill their goals by using the business process. Now, as the organization evolves, we need a change in the process; for example, the organization has decided to add one more department. Therefore, we have to update the associated business processes.

Technically, in jBPM, we cannot have an update in an already deployed process definition; we need to have a workaround. jBPM suggests three strategies for a process migration.

- **Proceed**: We will introduce the new process definition and retire the old definition. Retiring should be taken care of by the application so that all process instance calls for the process are redirected to the new process definition.

- **Abort**: The existing process is aborted, and we can restart the process instance with the updated process definition. We have to be very careful in this approach if the changes are not compatible with the state of the process instances. This can show abrupt behaviors depending on how complex your process definition is.

- **Transfer**: The process instance is migrated to the new process definition; that is, the states of the process instance and instances of activity should be mapped. The jBPM out-of-the-box support provides a generic process upgrade API, which can be used as an example.

These strategies can be discussed in detail and illustrated in the following chapters.

Summary

This chapter would have given you the "Hello world" hands-on experience in jBPM. With your jBPM installation ready, we can now dive deep into the details of the functional components of jBPM.

With this chapter, we have come to an end to the introductory chapters of *Mastering jBPM*. In the upcoming chapters, we will discuss the process designer, operation management, and core engine architecture in detail.

3
Working with the Process Designer

In the previous chapters, we were introduced to the concept of BPM and we got an overview of the jBPM tool stack and the family along some hands-on experience to build our first application using jBPM. The next four chapters will detail each jBPM component that we have discussed so far and the remaining chapters will guide you to customizing jBPM, integrating with the application architecture, and deploying a production-ready BPM application.

The first step that a business analyst with requirements for an application in hand would be the design and modeling of the business processes. This step is not often straightforward and involves multiple artifacts apart from the business process flow itself. This chapter will take us through the journey of designing a business process with all its etiquettes and will simulate the business process to understand how the designed business process would behave in a runtime environment.

The chapter guides you through the following:

- Business process modeling using BPM
- Data object modeling
- Using scripting and logic within BPM activities
- Modeling user interface forms that are attached to human activities in order to take input from human actors
- Simulating a business process to understand its runtime characteristics
- Using web-based and Eclipse tooling to work on the same project, and to collaborate between business users and technologists

We will discuss how to do the above primarily by using the web-based process designer. Eclipse-based tooling is also discussed but briefly, highlighting only the differences in performing modeling.

Web-based tooling

Web-based tooling for jBPM is an extension to the user interface framework provided for the Drools Guvnor knowledge repository. The framework provides a platform to create, maintain, and run multiple knowledge assets for a knowledge-based application. The jBPM extension (which is named jBPM-WB, the short form for jBPM workbench) uses the features of the platform to create and maintain applications and leverage the capabilities of the platform to provide user interactions with knowledge assets corresponding to the business process. This kind of abstraction helps the jBPM workbench to easily integrate the assets of the other software in the family and create an integrated development environment for the business logic integration platform.

In this section, we will focus on the asset editors for the knowledge assets that are part of jBPM, namely business process definitions, process and task forms, and data models.

Process modeling

The process designer provided with jBPM is an integrated environment to design, validate, and simulate business processes. The following screenshot highlights six distinct parts of the process designer:

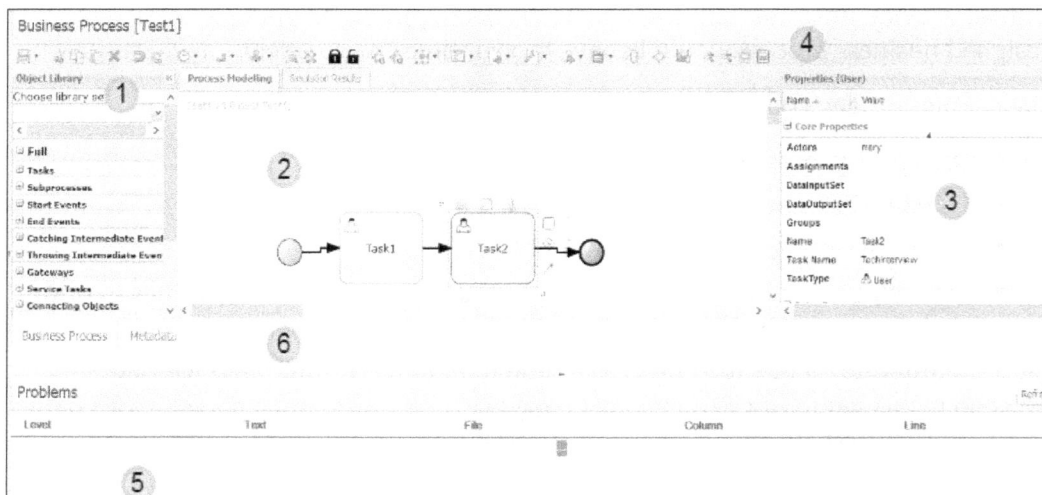

Parts of jBPM process designer

They are as follows:

- **1**: Object library
- **2**: Process canvas
- **3**: Properties editor
- **4**: Toolbar
- **5**: Problem visualization screen
- **6**: Metadata

Each of these parts is detailed in the following subsections.

The Object Library

The **Object Library** is a palette of constructs that are used to build business processes. The **Object Library** holds largely the BPMN-compliant constructs. The library is arranged as a drop-down menu with the categories of constructs serving as the headings. Users creating a business process can choose a construct from the palette and drop the construct in the process canvas. The following screenshot shows the default object:

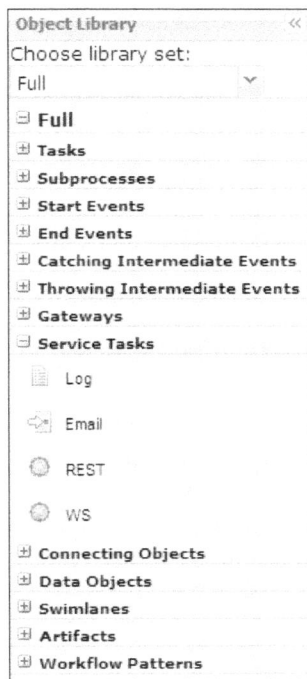

Apart from the BPMN constructs, the object repository also holds the workflow patterns. We have discussed these workflow patterns in *Chapter 1, Business Process Modeling – Bridging Business and Technology*, and they are solution templates for frequently occurring process design scenarios.

A developer can choose to customize the object library by adding extended and customized tasks; we can discuss these in the chapter exclusively for customization.

The process canvas

The process canvas is our workspace; we create the business process in this canvas by dragging and dropping the BPMN constructs provided by the object library, connecting and customizing them to create the desired business process. The process designer helps us to create a technically deployable business process and to model the aesthetics of diagramming by ordering and placing the objects, resizing the objects, and changing the color patterns. The following screenshot shows a sample business process designed using the process designer:

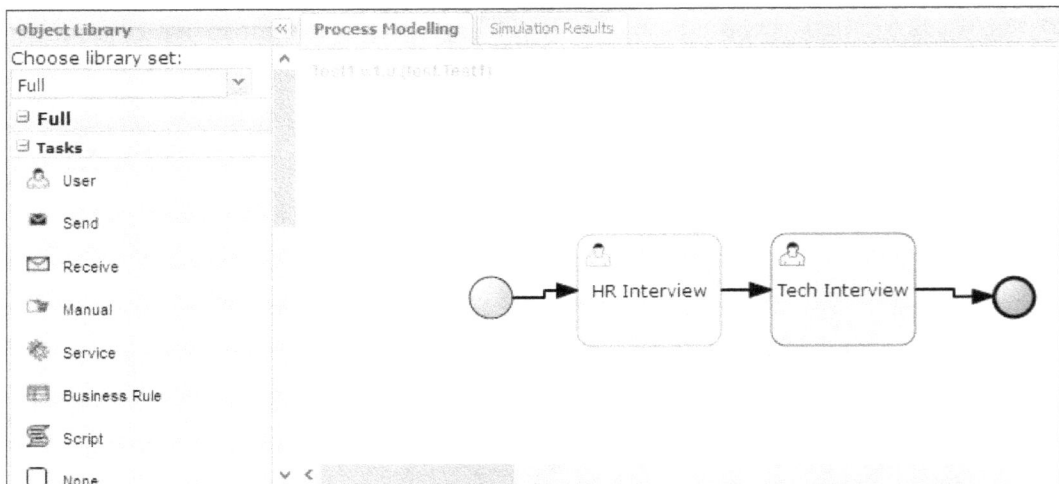

A useful tool, the community calls the morphing menu a great feature available in the process canvas. Once you drag and drop and select the object, we can see a tool menu around the object. The tools available in the menu will be different for different objects. There are three menus, one each at the top, bottom, and right of the object; they have a distinct collection of tools.

The right part of the menu has tools that help the user to work primarily on the process without always relying on the drag and drop feature of the object library. The user can add a task or gateway object to the canvas, include the connector, add an attachment, or add a data object. The following screenshot highlights the right part of the menu.

The top part of the menu has utilities for the following:

- Adding the task to the process dictionary (this will be discussed in detail in the subsequent section of the chapter)

- Viewing the **Node Source**, which helps in viewing the BPMN source of the object

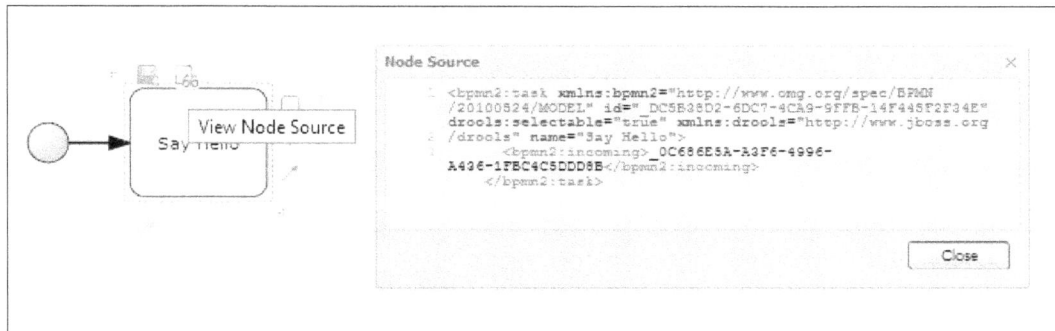

The preceding screenshot shows the top menu highlighted and a view of the **Node Source** window, which shows the source of the task object.

The bottom menu contains the tooling for converting a BPMN object into a similar counterpart. For example, we have placed a service task and have to change it to a human task. We will have to delete the task, replace it with another task, and restore the connections to it. Using this feature, and upon clicking the tool icon, we get a list of objects that we can interchange, and by selecting another object, we can change the object. This is a handy tool, particularly while maintaining a very large business process flow.

The following image shows the menu of interchangeable objects on a task element:

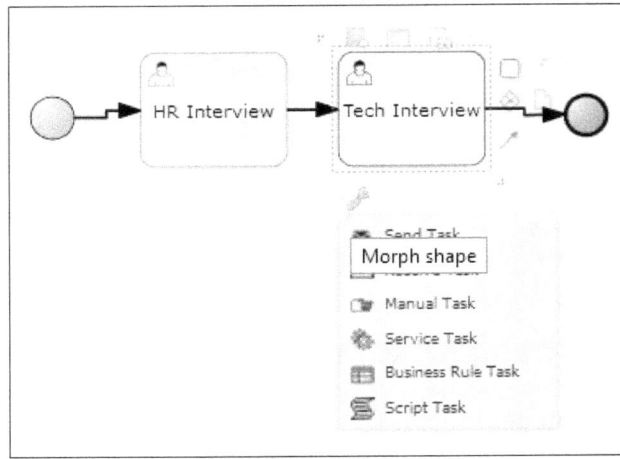

Properties editor

Apart from the visible parts of a business process diagram, each object in the business process has properties that can be customized. For example, in a human task, we have to assign an actor or a group. The **Properties** editor helps the designer to set these details. The **Properties** editor panel for each object can be obtained by selecting the object. For setting the process level properties, we have to click on the canvas itself.

The Properties editor panel has the following four sections:

- **Core Properties**: The Core properties section contains the properties that are essential for a particular BPMN element.

- **Extra Properties**: Extra properties as the name indicates are properties other than the core properties that are non-mandatory. For example, documentation for a task.

- **Graphical Settings**: Properties that can be changed for improving the aesthetics of the business process. The designers will have the flexibility to change the background, border, font colors, and the font size.

- **Simulation Properties**: Simulation properties are settings needed for creating the process simulation. The details of these properties are discussed in the *Process Simulation* section.

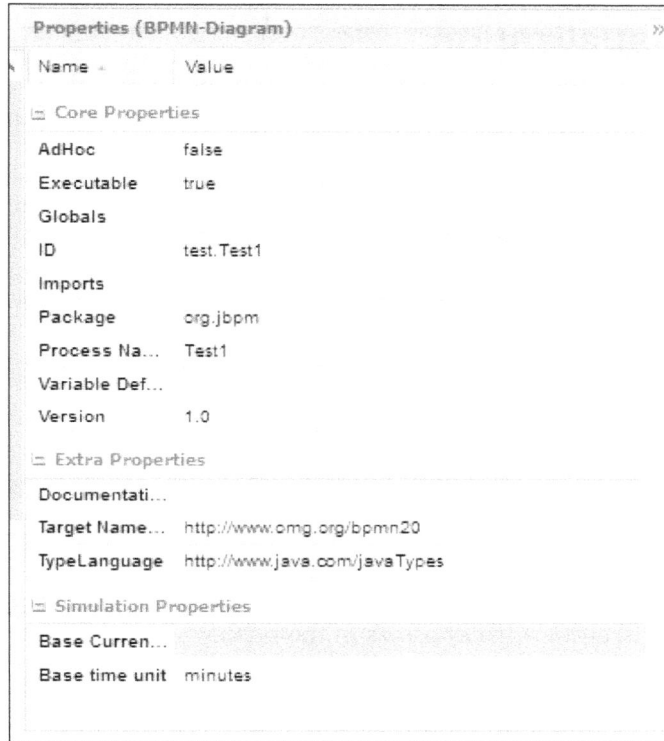

Properties (BPMN-Diagram)

Name	Value
Core Properties	
AdHoc	false
Executable	true
Globals	
ID	test.Test1
Imports	
Package	org.jbpm
Process Na...	Test1
Variable Def...	
Version	1.0
Extra Properties	
Documentati...	
Target Name...	http://www.omg.org/bpmn20
TypeLanguage	http://www.java.com/javaTypes
Simulation Properties	
Base Curren...	
Base time unit	minutes

The preceding screenshot shows the property editor for a process, showing core, extra, and simulation properties for a BPMN process.

The core and extra properties for each BPMN construct and the purpose are detailed in *Chapter 5*, *BPMN Constructs*.

Toolbar

Toolbar contains a set of utilities that aid in creating and maintaining a business process. Most of the utilities are common to any diagram editor (such as cut, paste, save, rename, delete, and zoom) and provide advanced features such as form modeling and process simulation.

Most of the tools are easily understandable from their names itself and may not need a detailed discussion. We will discuss the specific functionalities that need attention in the following sections.

Problem visualization

As we continue designing a business process, it is important to be informed about the syntactic and semantic errors in our business process. The problems section shows these errors in our application.

The following screenshot shows the errors from a BPMN process that has just started; that is, it contains only the start node. The errors indicate that the start node has no outgoing connection and the process does not have an end node:

Problems					Refresh ✕ ▾
Level	Text	File	Column	Line	
⊗	Process " [test Hello]: Start node " [1] has no outgoing connection.	Hello.bpmn2	0	-1	
⊗	Process " [test Hello]: Process has no end node.	Hello.bpmn2	0	-1	

Another provision included for the users is to visualize a problem in a specific process. This provision switches the editor into the validation mode. It can be accessed from the toolbar (refer to the following screenshot), and clicking on **Start validating** will show a list of errors in the process:

Also, if the error is in a particular object, the object is highlighted. Click on the highlighted object to get the list of errors.

The designer shows these errors by validating the business processes against a set of preconfigured rules; these rules are customizable. (Please see *Chapter 7, Customizing and Extending jBPM* for the customization techniques).

The Metadata tab

The **Metadata** tab shows the details of the business process as an artifact. It contains information such as the user who created and modified the artifact, the format of the artifact, and so on. Also, there is a provision to have a discussion about the artifact, which will be very useful in scenarios where we have a larger distributed team of business analysts who create and manage the business processes.

Data object modeling

A part of business domain modeling is to identify all the entities in the problem domain, the relationships between the entities, and their attributes. The data associated with these entities are often collected, updated, and removed via business processes. Thus, these entities become a part of our business process.

For example, **Customer** is a data entity that occurs in almost every business domain, and customer creation (the process of adding a customer) is a very common business process. JBPM provides data modeling features where these entities can be created as data objects and be included in the business process.

Let us discuss data modeling with an example. We consider a very primitive customer creation process with a human task to capture customer details and a service task to call the customer creation service available. The image given below shows the business process under discussion.

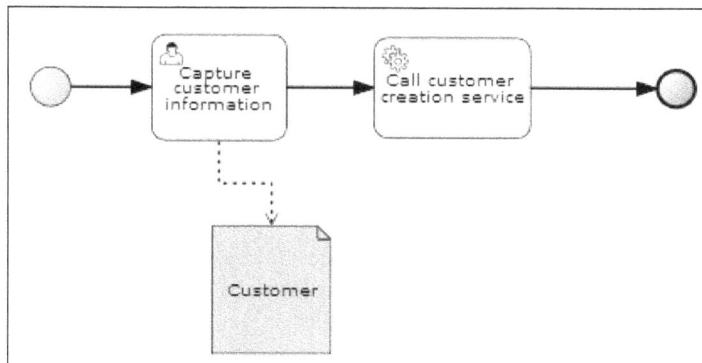

To create the `Customer` data object by using the process designer, follow these steps:

1. Log in to the workbench.

2. Go to **Authoring** | **Project Authoring** | **Tools** | **Data Modeler** and click the **Create** button. The following screen pops up:

3. For creating the customer data object, give an identifier, a label, and the package name for the unique identification of this data object.

Well, as you must have noticed, there is another (optional) field called **Superclass**. This needs a bit of explanation on the technical side of how the data object is considered in the design of jBPM.

The data object modeled by us is created as a Java object with the attributes specified by us and included in the application as its dependency. The superclass points to the inheritance feature of Java, and we can use this feature in scenarios where data objects have parent–child relationships.

After creating the data object, we can use the **Create new field** provision (see the following screenshot) to add the attributes corresponding to the data object. The following screenshot shows the **name**, **age**, and **sex** attributes added to the **Customer** data object:

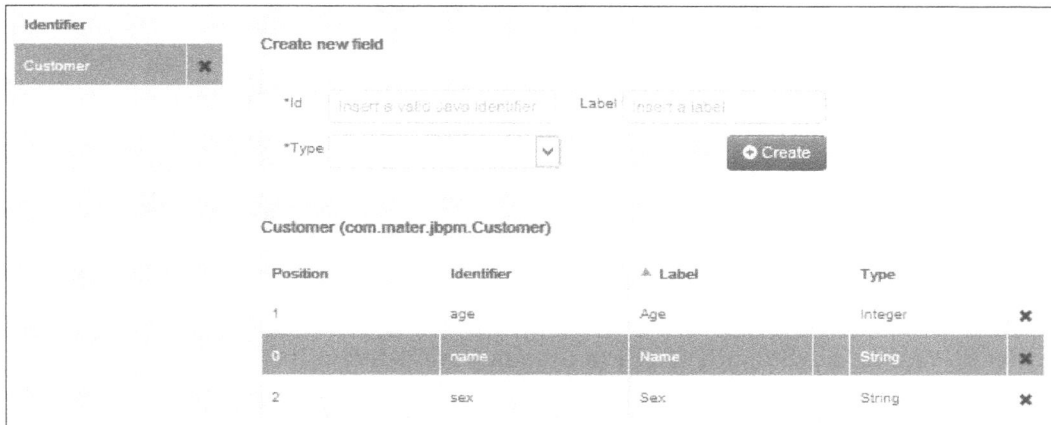

Save the data object. Now, we have to utilize the data object in the business process.

In the business process, we have a human task that captures the information of a customer as its input and maps it to the output set of the task. For doing so, in the process designer, add the customer objects as **Type** in the input and output set of the human tasks.

The following screenshot shows the property editor for the **Capture customer details** human task, where the customer data object is used:

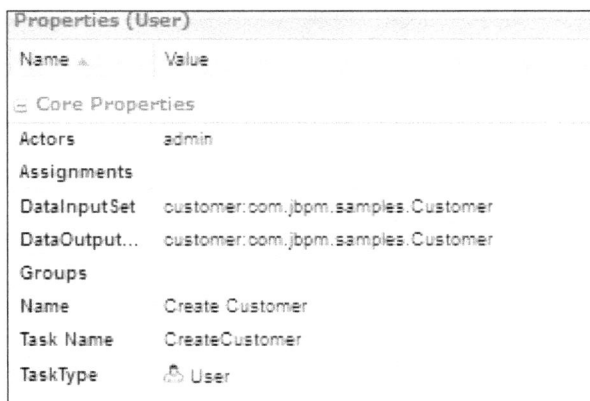

Form modeling

Continuing the discussion with our example of primitive customer creation, the next problem will be how to provide a user interface for the human task (capturing customer information). Form modeling provides a solution for this scenario. A business analyst can design a form that can be used to capture the information attached to a task.

To be precise, the data collected through the form is mapped to the data output set of the human task. When we run the **Create customer** business process, the "Capturing customer information" task is assigned to an actor (or a group) and will be available in the inbox. For completing the task, the user will have to enter the details into a user interface that the engine renders by using the model of the form.

Creating a form

So, let us explore how to model a form for a task included in the business process.

1. In the **Create customer** process, select the **Capturing customer information** task.

2. In the highlighted tool menu, the top part contains the tool for editing the associated form. Click on the tool, and select graphic modeling; this takes us to the form modeling screen:

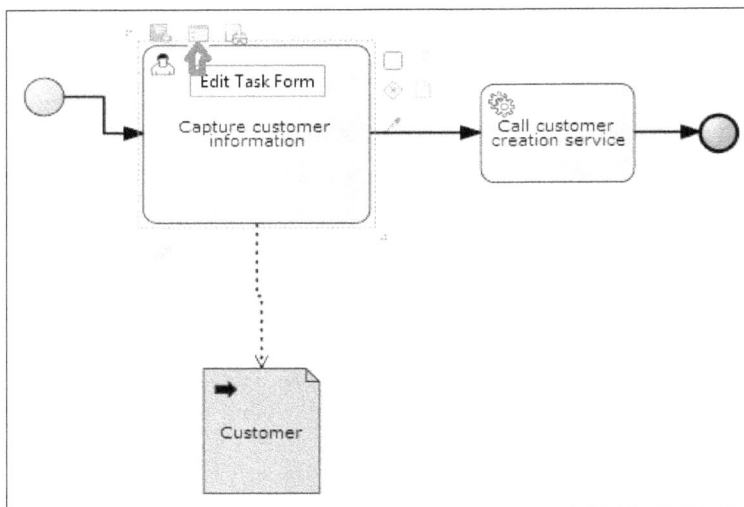

For modeling the form, first, we have to create the data origins, where we define the mapping of data input sets to form variables and the form variables to data output sets. In this scenario, we have to map the data input and the data output of the customer variable.

We have to fill the **Id, Input Id**, and **Output Id** fields and choose the render color for the form. Further, we have to select the data models that the form has to represent and click **Add data holder** to add a data model as a data origin for the form.

The following screenshot shows the form modeling screen after adding the customer data origin:

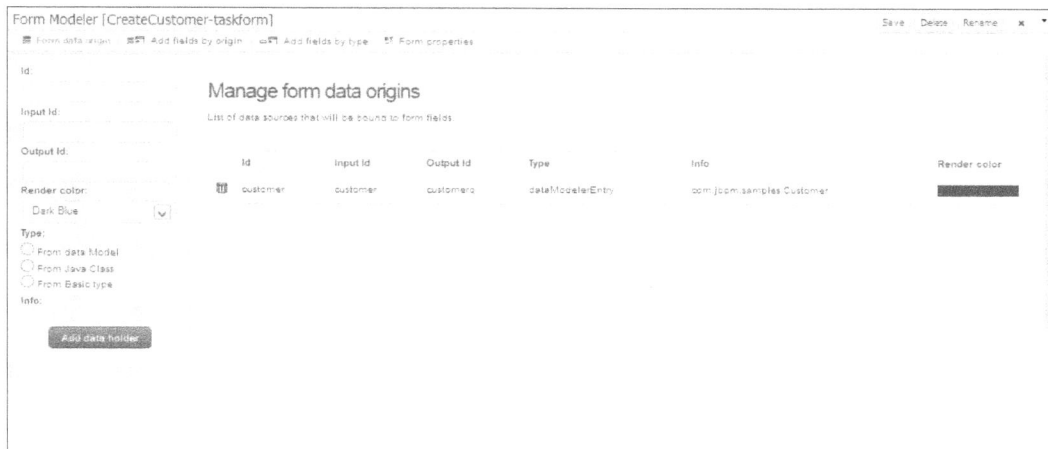

Now, we can model the look and feel of the form. For this, we choose the **Add fields by origin** tab, where the data origins are listed and add to the form canvas the fields attached in the customer data object (as shown in the following screenshot).

Select a field, use the toolbar to edit the field properties, and move the field around in the form, to improve the aesthetics.

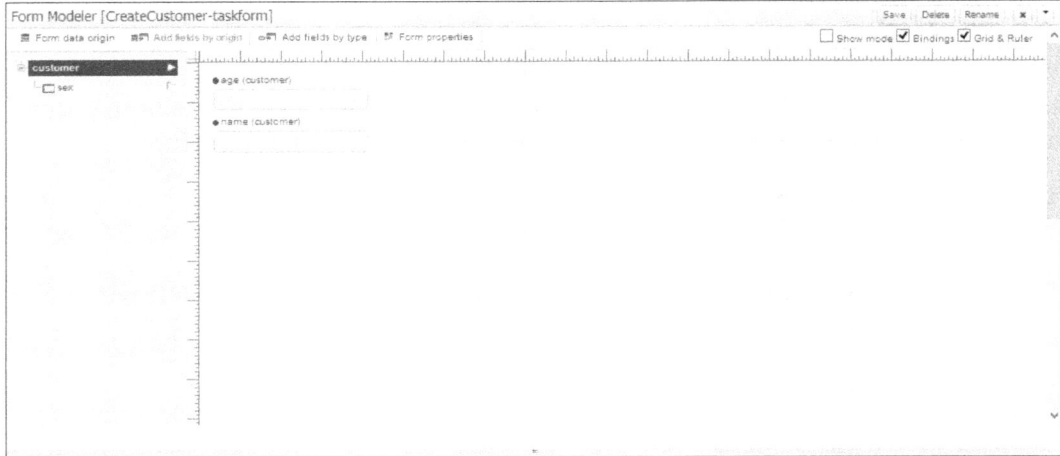

Setting the field properties

For each field included in the form, we can specify a set of properties that determine its runtime behavior and its layout.

For using the **Properties** panel (as shown in the following screenshot), select the field, and on the tool menu, click the edit button.

The properties include the following:

- Properties for validating data integrity
- Size of the input text
- Maximum characters supported
- Mandatory or not
- Read-only or not
- Password field or not
- Pattern of the data expressed as a regular expression
- Properties for specifying the look and feel
- CSS style of the label and the field
- Type of the field if the type of data has multiple user interfaces; for example, string can be capture in a field or text area or by using a rich-text editor

- Properties for data binding
- Properties for increasing usability
- Setting default values
- Inferring values using formulas

The following screenshot shows the **Properties** editor for the string type data:

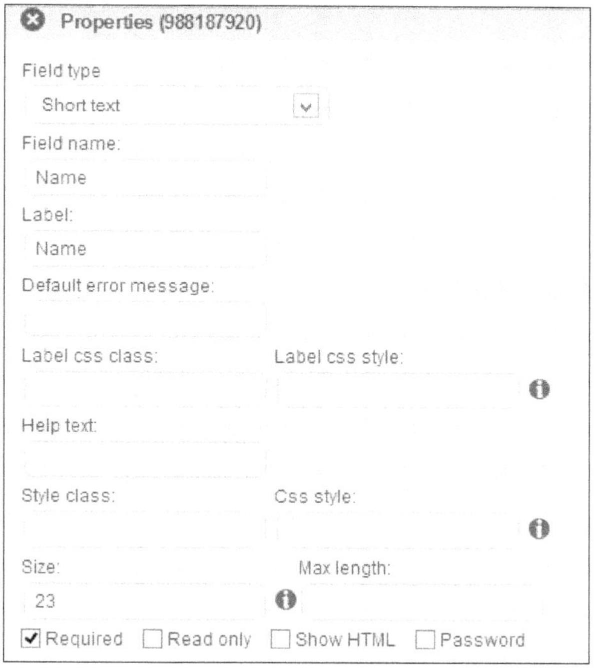

Setting a default value

A general usability and productivity improvement technique in data entry screens is to provide default values for the fields. The form modeler supports this feature by using Java and **XPath** expressions.

> XPath is a query language that can be used for selecting nodes in an XML document. See http://en.wikipedia.org/wiki/XPath.

Suppose that in the customer creation process example, we have to default the field **sex** to **Male**, assuming that there are more male customers in the problem domain than female customers. So, in the **Default value formula** field of the property editor, we can use the expression as given in the following screenshot:

Default value formula:

= "Male"

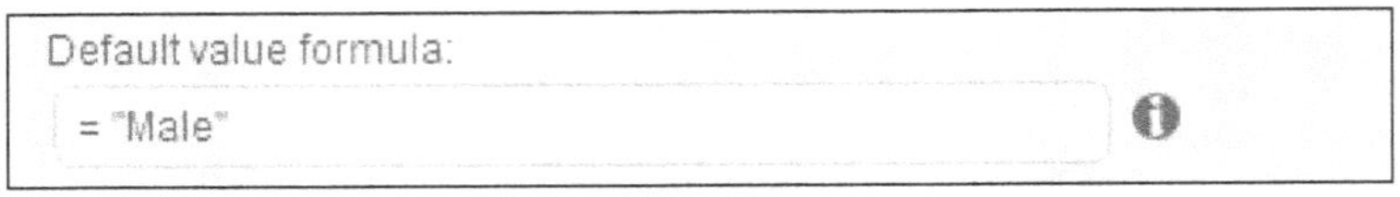

Another common occurring scenario is to default a date field to the current date.

Default value formula:

= new java.util.Date()

Inferring field values

Another feature that we require in a data entry screen is to populate the values of some fields on the basis of the values of the other fields. For example, if we are capturing the information of a product (say, a laptop), there will be multiple cost components such as price, tax value, and discount. The total cost of the product is derived from these three components. So, we can set the formula for calculating the total cost by using an XPath expression referring to the fields for price, tax value, and discount (see the following screenshot).

☐ Required ☑ Read only
Formula:

={unit_price}+{tax_rate}-{discount_rate}

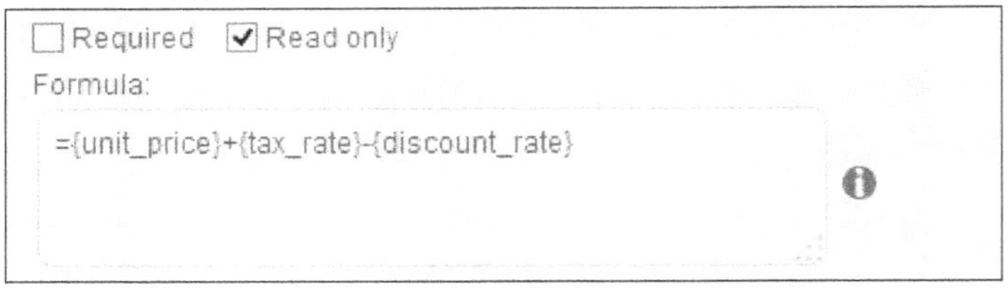

Making the field read-only would complete the trick.

Subforms

In a form, it is common to contain multiple sections. For example, while capturing customer information, we also need to capture the address information. Further, there are two type of address: one is the permanent address and the other is the communication address. Considering reusability in the form design, the most elegant way of modeling this scenario will be to create a single address form and including it twice in the **Customer** form.

Form modeler supports this scenario by using a feature called subform, where we can include one form in another. For achieving this, we have to create **Address form** and we have to go to the **Add field by type** tab of **Customer form** and add **Simple Subform** and choose the property of **Default form** as the already created **Address.form**:

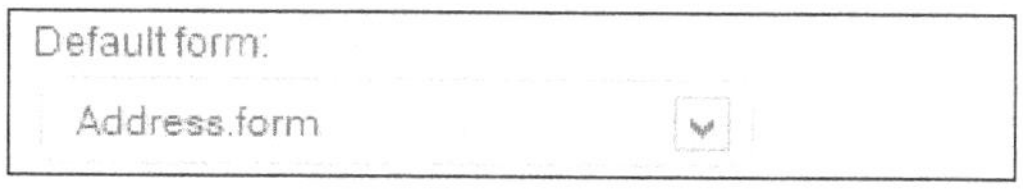

Multiple subforms

Yet another requirement is to have multiple objects of the same item in a form. Consider a scenario of the order management process. We have to capture the order information. An order typically consists of two parts: one is the order itself having information such as who the requestor is and at what time the request was made, and the other part is the requested item that consists of the product and the quantity ordered. This can be multiple; that is, the requestor may have multiple items in an order. A typical form to capture these details will be as given in the following screenshot:

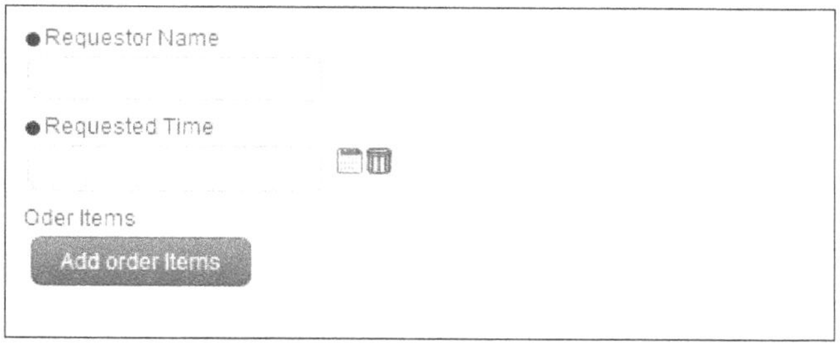

Furthermore, on clicking **Add order Items**, we will get a row of order items.

How do we model this scenario in the form modeler?

Follow the same method as that for the subform. Create data models for Order and Item. Create forms for Order and Item. In **Order form**, use the **Multiple subform** form type.

If we take the property editor of **multiple subform**, we will get a lot of options to improve the look and feel of the form, to control what operations can be seen in the form, and so on. Play around and get a good feel of it.

Process simulation

So far, we have discussed a lot about business process modeling and aids that jBPM provides for it. The modeled process can now be deployed to know the runtime characteristics. Wait! Don't we have to analyze and verify the characteristics of the business process? JBPM provides tooling for analyzing the runtime characteristics of a modeled business process, and this is called business process simulation.

Process simulation helps us to do the following:

- Optimize the business process design by pre-execution
- Understand the resource utilization of human actors involved in the business process
- Understand the performance characteristics of the business process by prediction and analysis
- Continuously improve the business process design by minimizing the errors caused by a change

We can progress with the discussion on process simulation by using a sample business process flow as shown in the following image. The process flow depicts a business process for an online transaction, including the collection of customer information, sending the one-time password, and validation of the password. The consideration for choosing this process design for illustration was to have multiple process paths:

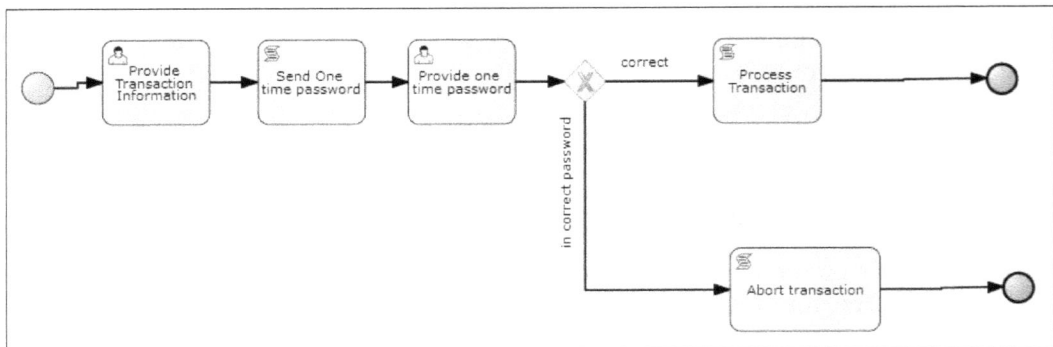

Process paths

The process paths feature helps the process designer to see all the possible combinations of paths in the business process. This will help us to understand any flaw in the logic we have used.

We can access the **Process Paths** tooling from the designer toolbar, as shown in the following screenshot:

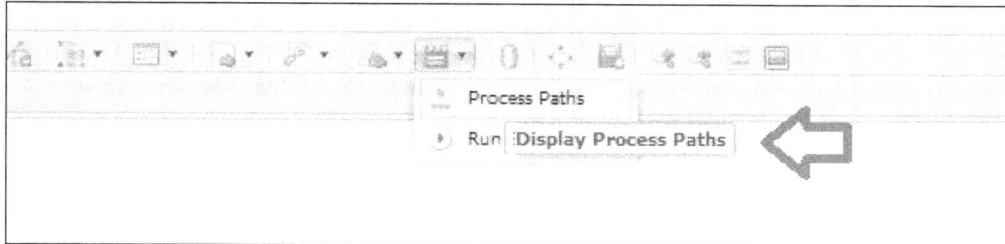

Select the **Process Paths** menu; it will calculate and list the paths that the process flow can possibly take. Select one of the paths, and click **Show Path** to highlight it in the process canvas:

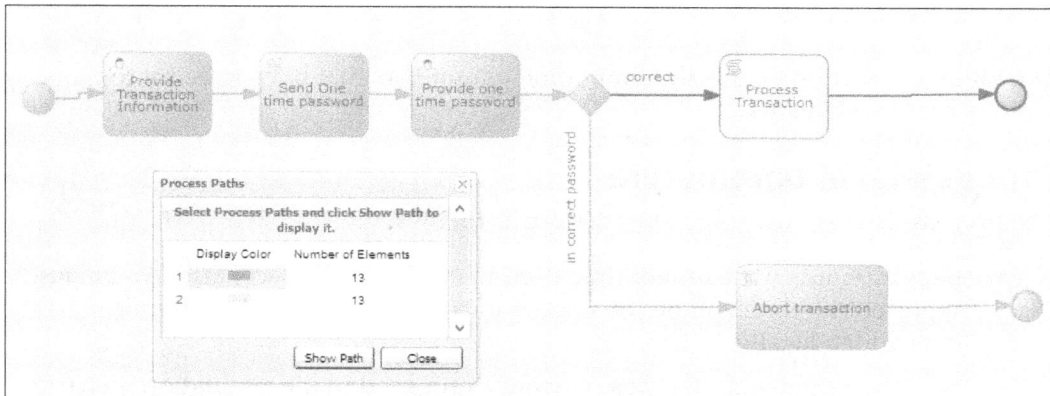

The preceding screenshot shows the process with a path highlighted. This tool becomes handy, particularly while designing complex business processes.

Simulation parameters

The process simulation engine needs some parameter inputs other than the information available in the business process; these parameters collectively describe or define the scenario of business process simulation. The parameters to be captured for scenario creation vary with the type of business process elements in a business process. For example, we need to capture the working hours with the human tasks for simulating resource utilization.

Simulation properties can be changed using the property editor and vary when selecting different process elements. The following screenshot shows the property editor showing the simulation properties for a human task (**Provide one-time password**):

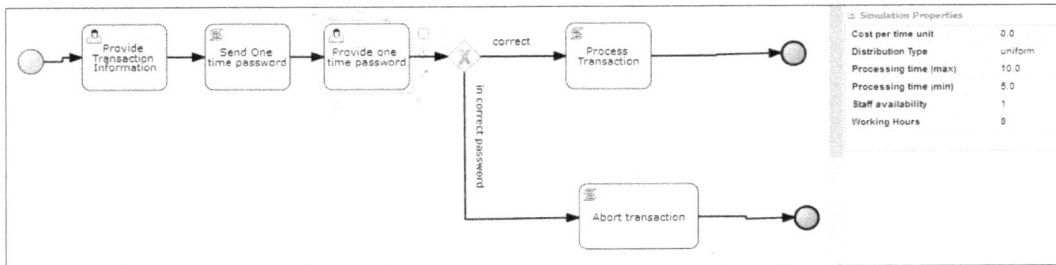

The following section details the simulation parameters that have to be set for various process elements.

Process-level parameters

The process-level simulation properties are as follows:

- `baseTimeunit`: The unit of time used in the simulation scenario. All values representing time will be considered to be expressed in this unit unless overridden locally.

- `baseCurrencyUnit`: The base currency unit used in the simulation scenario expressed using the ISO 4217 (three-letter code) standard. For example, the US dollar is represented as **USD**. Similar to baseTimeUnit, all values representing currency will be considered to be expressed in this unit unless overridden locally.

> ISO 4217 is an international standard, established for the purpose of having internationally recognized codes for the representation of currencies.

Task-level parameters

The task-level simulation properties are as follows:

- `unitCost` (labeled as cost per time unit): This is the cost per time unit that has to be paid for performing the task. The cost is expressed as the number of `baseCurrencyUnit` properties per `baseTimeUnit`, and the default value is zero.

- `distributionType`: This specifies the statistical distribution of the processing time of this task over the period of execution of multiple instances of the business process in the scenario.

The supported statistical distribution types in jBPM are as follows:

- **Uniform distribution**: Uniform distribution or rectangular distribution has equal probability for all values between the minimum and the maximum processing time values

- **Normal distribution**: It is one of the common distributions in nature represented as a symmetric bell-shaped curve and is specified by the mean and standard deviation values of the processing time

- **Poison distribution**: It is used to estimate the number of arrivals within a given time and is specified by the mean processing time

> For in-depth understanding of the distribution type, please try understanding probability distributions; see http://en.wikipedia.org/wiki/Probability_distribution.

- `staffAvailability`: The quantity of resources available for the task and is set to a default of 1. This parameter is applicable only to human tasks.

- `workingHours`: The working hours for a human resource. This parameter is also applicable only to manual (human) tasks.

Flow element parameters

The simulation property attached to a sequence flow is *probability*, which is the probability of the control being passed to this element. This is attached to the sequence flow process element. If we specify a probability value of 50 percent to each of the two sequence flows outgoing from a gateway, it means that the two sequence flows are equally probable to occur.

Running simulation

Now, as we have discussed about setting the simulation properties, we can run the simulation. The menu is available along with **Process Paths** in the process designer toolbar:

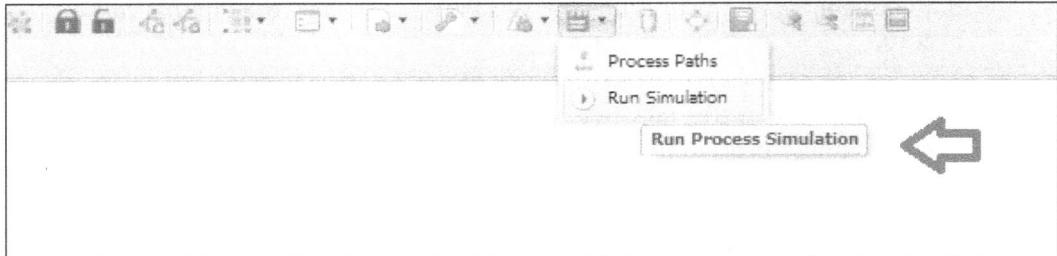

We will be prompted to provide details such as the number of instances to be simulated and the interval between process simulations. The simulation process is asynchronous, and once completed, the simulation tab will be populated with the simulation results.

Simulation results

The summary of the simulation is available on the right hand side of the simulation tab. This contains the information of the simulation and a hierarchy for navigating through the simulation results at different levels such as process, process elements, and paths. The following screenshot shows the summary information for the sample process:

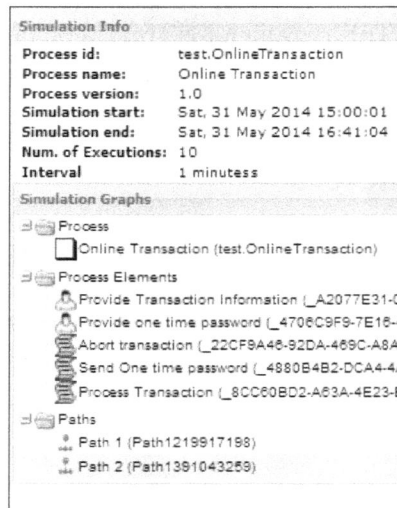

The results section is quite fleshy, with different types of charts to choose from that represent data in multiple ways. Play around with the options there. The following are the major datasets included in the simulation results:

Process simulation results: If we select the process (from the hierarchy shown) to view the execution result, we can view the following:

- **Execution times results**: This shows the maximum, minimum, and average time of execution.

- **Activity instances**: This shows the activity (task) instances created during the simulation.

- **Total cost**: Shows the minimum, maximum, and average value of the projected cost for the business process.

- The following image shows the execution time results of the online transaction business process:

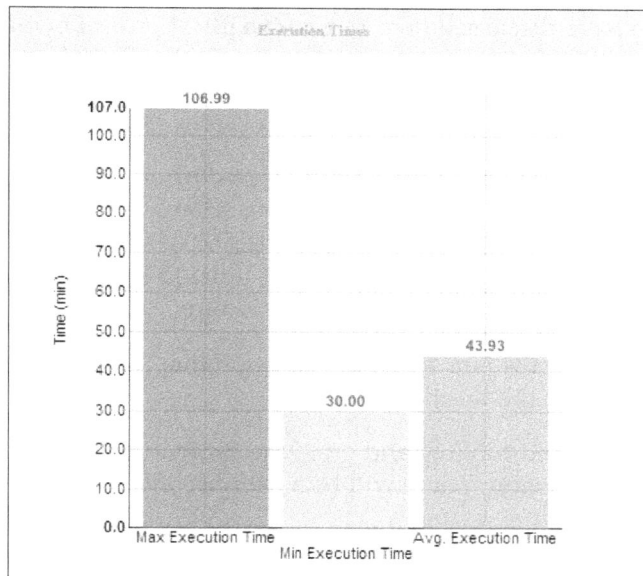

- **Process element-wise simulation results**: Select each process element in the hierarchy to show the process element-wise simulation results. For service tasks, this basically shows the minimum, maximum, and average values of the execution time. Further, for human tasks, it also shows the resource allocation details, wait time for resources, and the cost factor.

- **Path simulation results**: Path simulation results show the number of times that the path has been executed and its percentage contribution.

Eclipse-based tooling

In the previous sections of this chapter, we have concentrated on modeling the process and utilities available in the web-based tooling provided by jBPM. The web-based tooling primarily focuses on the business analysts. Can the business analysts create an application on their own? We must acknowledge that we are not yet there, but, of course, there are only a few gaps left to reach that destiny. On today's course, we need the help of technologists to create a completely functional application.

jBPM tooling considers this reality in the form of the updated Eclipse tooling that helps developers to create a jBPM application or collaborate with the team of business analysts to complete an application.

Importing a project created using web-based tooling

In this chapter, we have discussed how to create a jBPM project by using Eclipse tooling. Here, we will discuss how to import a project that was created in the process designer (possibly by someone in the business analyst role) into Eclipse and continue working on it.

The default implementation of the knowledge repository in jBPM utilizes Git, a widely used source code management system. So, when we create a project by using web-based tooling, it actually creates a project in Git. Git supports people to work concurrently on a project, and this facility can be utilized for collaborative working between users of web-based tooling and Eclipse tooling.

So, first, we should know the Git repository location that jBPM uses as the knowledge repository for the sample project:

1. Log into web-based tooling, and go to the **Project Authoring | Administration** menu. There will be a list of repositories and projects.

2. Select the project we want and we can see the Git repository URL. There will be two URLs available: one using the **git** protocol and the other for the **ssh** protocol. The screenshot of this section is as follows:

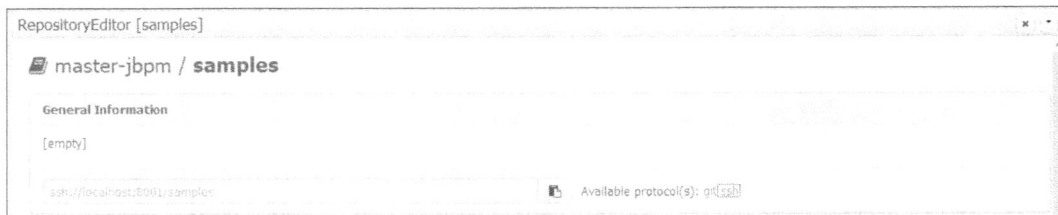

RepositoryEditor [samples]

master-jbpm / **samples**

General Information

[empty]

ssh://localhost:8001/samples

Available protocol(s): git ssh

3. Copy the URL using the tooling available.

4. Open Eclipse (that is installed with the jBPM installation).

5. Go to **File | Import | Projects from Git**, click **Next**, select **URL**, and click **Next**. We will get the window shown in the following screenshot.

6. Enter the ssh URL and authentication details:

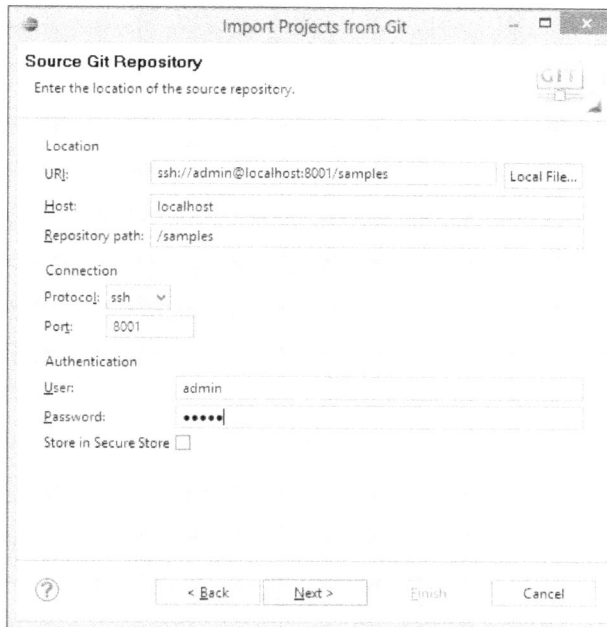

7. Proceed with the wizard and finish importing the project; in Eclipse **Package Explorer**, we will get the project imported. The project structure is the Maven project structure for a Java project. Explore and we can see the data object, forms, and business processes that we have created:

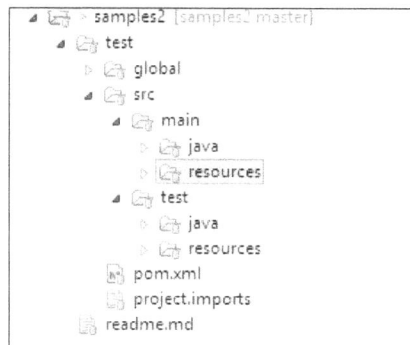

Process modeling

Eclipse-based tooling comes with a BPMN diagram editor for modeling business processes. The editor provides similar features as the web-based process designer. It consists of a canvas where we can visually illustrate the business process, **Palette**, which acts the BPMN object library, and the property editor for setting the properties of each BPMN element. The following screenshot provides the BPMN diagram editor:

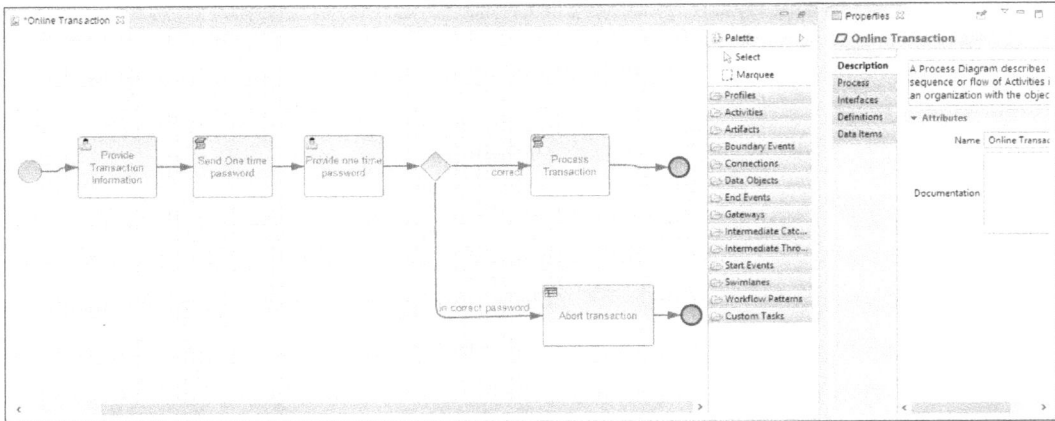

We are not discussing the features of the editor in detail as they are similar to the features explained for web-based tooling.

Data object modeling

For data object modeling, jBPM doesn't provide any visual modeling tooling, but the source code generated internally while creating a data object is provided to the developer user. This source code is in Java, and the objects are represented as **Plain Old Java Object** (**POJO**) and decorated using annotations from the knowledge API.

The following screenshot shows `Order.java` created for the `Order` data object created in the process designer:

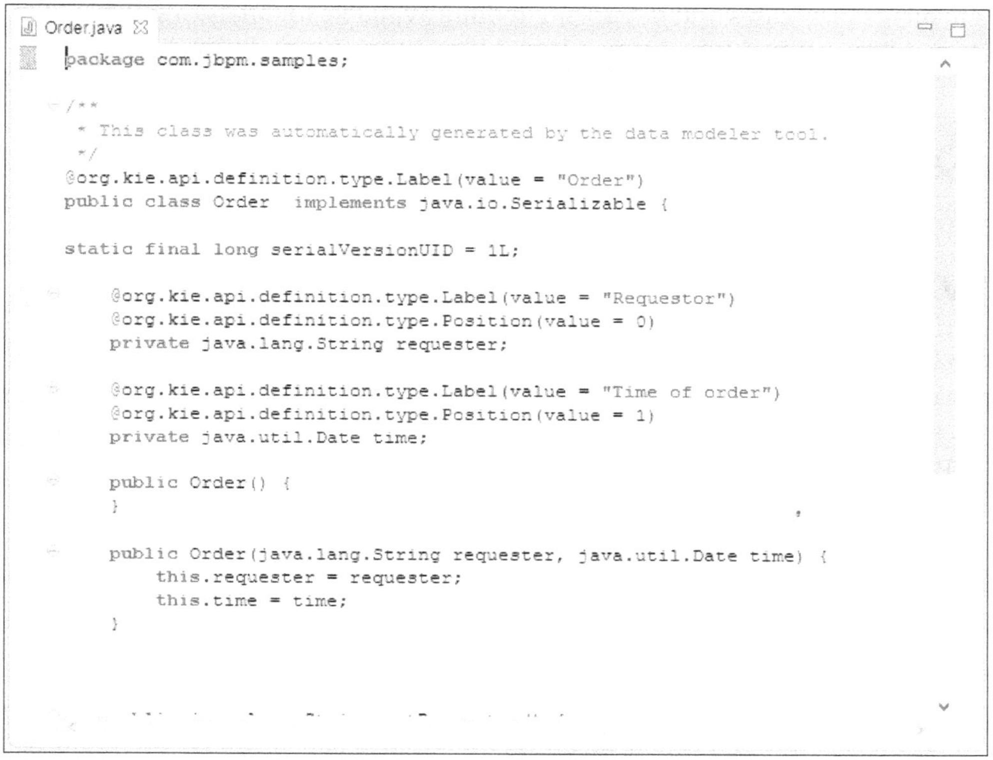

```
Order.java ⊠
    package com.jbpm.samples;

    /**
     * This class was automatically generated by the data modeler tool.
     */
    @org.kie.api.definition.type.Label(value = "Order")
    public class Order  implements java.io.Serializable {

    static final long serialVersionUID = 1L;

        @org.kie.api.definition.type.Label(value = "Requestor")
        @org.kie.api.definition.type.Position(value = 0)
        private java.lang.String requester;

        @org.kie.api.definition.type.Label(value = "Time of order")
        @org.kie.api.definition.type.Position(value = 1)
        private java.util.Date time;

        public Order() {
        }

        public Order(java.lang.String requester, java.util.Date time) {
            this.requester = requester;
            this.time = time;
        }
```

Form modeling

Similar to data object modeling, developer tooling doesn't include any visual tooling for form modeling but is available as a raw file that can be hacked by developers.

The files with the `.form` extension contain the properties within the form and the files with the `.ftl` (which refers to FreeMarker template files) extension contain the layout information of the form.

> For more details on FreeMarker templates, see `http://en.wikipedia.org/wiki/FreeMarker`.

Process simulation

There is no tooling as such available for process simulation, and the developers have to rely on unit test cases and debugging tooling available for analyzing the runtime characteristics of a process.

> The view for writing unit test cases and process instances is discussed in the *Writing automated test cases* section of *Chapter 2, Building Your First BPM Application*.

Saving changes to the knowledge repository

In web-based tooling, saving the artifact would be reflected in the knowledge repository. Eclipse-based tooling is a bit different in this aspect. The saved files will reflect only in our local file system; for synchronizing with the knowledge repository (here, the Git repository), we can use Eclipse tooling for commit and pushing to the Git repository.

1. Right click on the project (**Package Explorer**), and go to **Team | Commit**. It will take you to the commit screen shown in the following screenshot.

2. Select the artifacts to be moved to the repository, provide a commit message, and use the **Commit and Push** button to push the changes to the knowledge repository:

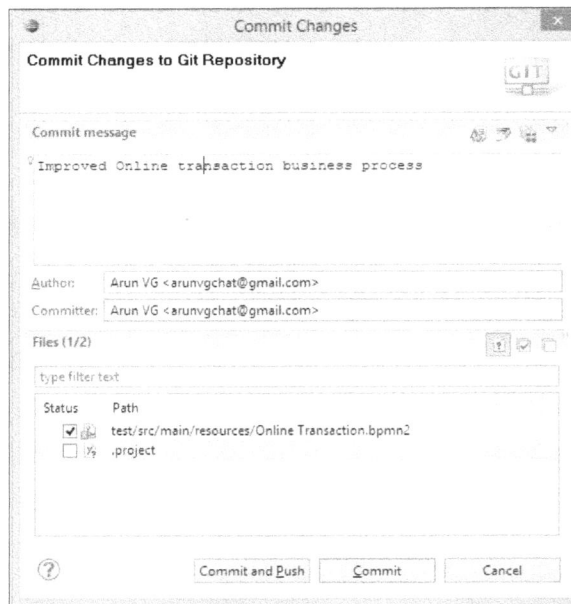

Summary

The chapter focused on the tooling available for process design and covered in detail the various features available in both web and Eclipse-based tooling, targeting business users and technologists, respectively. The chapter also helps in getting a hands-on experience of how both web and Eclipse-based tooling can be used collaboratively in process design.

Now, as we have discussed process design in considerable detail, let us explore the tooling available for operation management in the next chapter.

Also, *Chapter 5*, *BPMN Constructs*, can be considered an extension to process designing, where we will discuss in detail each BPMN construct that can be included in a business process.

4
Operation Management

This chapter will illustrate all the tasks that are required to perform jBPM operations by walking you through the following topics (focusing on the jBPM KIE workbench and related tools):

- jBPM environment configuration: Git and Maven repositories, organizational units, and user management with basic administration and permissions by **role-based access control (RBAC)**
- New jBPM asset management feature and module deployment
- Process and task management
- jBPM auditing and history log analysis with a working example of BAM
- Job and command scheduling with jBPM Executor

This chapter requires a working knowledge of both Git and Maven, which play a central role in the KIE workbench architecture. You will be asked to work with Git and to deploy artifacts to Maven. Let us start by reviewing the typical software architecture of a jBPM 6.2 development system with the aim to shed some light on the new system components and the way they interact.

An overview of the KIE workbench, Git, and Maven

At first, the jBPM 6.2 architecture may seem a bit hard to grasp, since several new components have been integrated to provide the developers with industry standard tools for making it easier to support source code management and building/deployment. The jBPM platform integrates with the Git and Maven repositories so that you can share Kie business assets and publish Kie modules to remote teams. Let us see how Git and Maven fit into the Kie platform (shown in the following image).

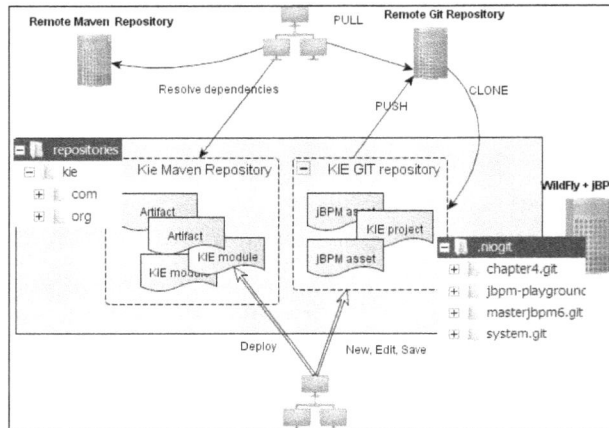

The KIE workbench manages assets from its Kie Git repositories (either brand new or cloned from remote repositories). All Kie Git repositories can be found in the .niogit folder. The Kie deployment process installs the modules into the Kie Maven repository (located in the repositories/kie folder). This repository is publicly accessible via either the Git or the SSH protocol.

Working with Git

The KIE workbench enables us to create a new empty Git bare repository or to clone a remote Git repository into a brand new Kie bare repository. However, the workbench does not allow us to import assets into an existing branch of a Kie repository.

> "Bare" repositories exist in Git as a way of having a central (mainly remote) repository that a number of people can push to. For details on the bare Git repository, please see the official Git documentation.

We can manage repositories from the **Authoring | Administration** menu (**Repository** item). Let us now put Kie repositories to work.

Cloning a remote repository

We are going to clone the `chapter 4-pizza` example repository from GitHub. It hosts two projects that we will use later in the chapter to experiment with the deployment process. To clone a remote repository, open the **Repositories | Clone Repository** dialog (see the following screenshot) and configure the parameters as follows:

- **Repository Name**: `chapter4-pizza`
- **Organizational Unit**: Use the default provided by jBPM or create a new one (This is not relevant at this stage; let us set its value to `demo`)
- **Git URL**: `https://github.com/masteringjbpm6/chapter4-pizza.git`

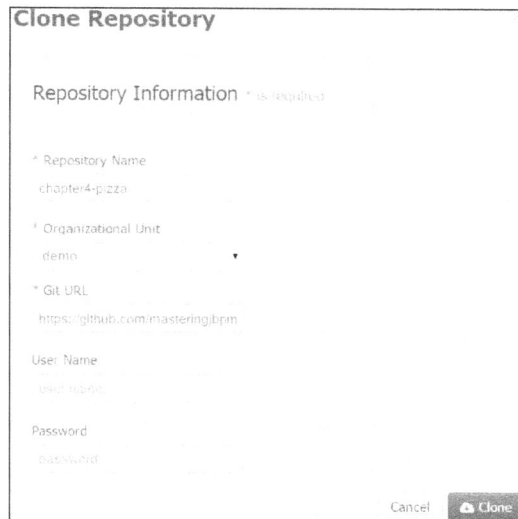

We have already said that both cloned and new repositories are placed in the `.niogit` folder that you can find in the KIE installation folder. In addition, all KIE repositories are shared by default at the following URL:

`git://localhost:9418/{reposname}` or `ssh://localhost:8001/{reposname}`

where `{reposname}` is the **Repository Name** that you provided in the dialog window (for example, `chapter4-pizza`).

[Never clone repositories from the KIE `.niogit` folder directly; always use the repository URL.]

The `.niogit` folder also contains the `system.git` Kie repository, which is used as a store for metadata and settings; we will look at it in an upcoming section.

Making changes and committing

Modifying the project assets from the KIE workbench means that your changes are going to be committed into the KIE Git repository. Let us edit the jBPM process definition and see what happens upon saving the asset. Open the **Authoring | Project Authoring** menu (**Project Explorer**) and change the path to `demo/chapter4-pizza/pizzadelivery`; you should have the **pizzadelivery** process listed under the **Business Processes** group (see the following screenshot).

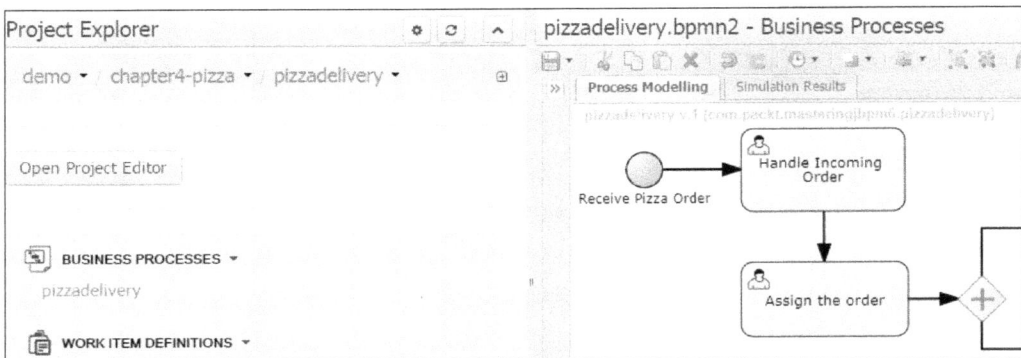

Just drag and move, for instance, the process start node, and then, select the **Save** menu item. The system prompts you with a **Commit** comment; enter `workbench editing` and press **OK**. Now, if we go and check the `.niogit/chapter4-pizza.git` repository and issue the following command:

```
git log
```

We can see the following two commits (the older commit was the one performed during the GitHub repository initial creation, and the other is your last commit):

```
commit 165a0b07f1b50962696640bcb79072458f4c82d4
Author: admin <>
Date:   Sun Mar 22 11:33:16 2015 +0100
```

```
workbench editing {/pizzadelivery/src/main/resources/pizzadelivery.
bpmn2}
commit a32f9a2a9bc835e74abcb78348878d8d2fc96140
Author: admin <fiorini.s@gmail.com>
Date:    Mon Mar 23 07:48:50 2015 +0100
    pizza commit
```

We can get the first commit files by specifying the commit ID:

```
git show --name-status -r a32f9a2a9bc835e74abcb78348878d8d2fc96140
Date:    Mon Mar 23 07:48:50 2015 +0100
    pizza commit
A        pizzadelivery/.classpath
A        pizzadelivery/.project
A        pizzadelivery/.settings/org.eclipse.bpmn2.modeler.core.prefs
```

Pushing to the remote repository

Let us now check the remote origin setting for our `chapter4` repository first by using the following command:

```
git remote -v
```

The following information is printed:

```
origin  https://github.com/masteringjbpm6/chapter4-pizza.git (fetch)
origin  https://github.com/masteringjbpm6/chapter4-pizza.git (push)
```

In order to update the remote origin GitHub branch (master), we issue the following push:

```
git push origin master
```

Since the repository is protected, you will be prompted for the repository username and password (we are using our account here):

```
Username for 'https://github.com': masteringjbpm6
Password for 'https://masteringjbpm6@github.com': *****
```

Removed for clarity…

```
To https://github.com/masteringjbpm6/chapter4-pizza.git
79bcf3a..36b57c1  master -> master
```

The remote branch is finally updated.

For fixes and enhancements, we encourage you to contribute to our example GitHub repository; in case you want to experiment with the example projects on your own, please follow the next section.

New repository

By cloning our GitHub repository into a KIE repository, you cannot (as you are not a contributor) send changes upstream. Apart from forking on GitHub, creating a new empty Kie repository is the right choice if you want KIE to centrally manage your projects. To do so, follow these steps:

1. Create a new KIE Git repository; remember to leave the **Managed Repository** option unchecked for the moment, since this turns your repository into a Maven aware repository making it hard to adjust it when projects are added from an external source (we are going to address it in the *Asset management example* paragraph).

2. Clone the remote GitHub repository from the command line, using your preferred Git client.

3. Change your Git working copy remote origin to the new Kie repository.

4. Commit and push to the Kie repository `master` branch.

Please see the *Git cloning and managed repository* section for a complete example of repository creation and project setup from GitHub.

jBPM 6.2 introduces a distinction between the plain Git repository (unmanaged) and a new kind of "smart" repository (managed) that we can create by setting the **Managed Repository** option in the **New Repository** dialog window.

Managing assets – managed repositories

As we have seen, managing Git project sources and Maven with KIE might be challenging: cloning, committing, setting repository remotes, pushing upstream, and so on, and we did not even consider Git branching in our example. jBPM 6.2 simplifies the way repositories and project source code are managed by introducing a new feature (asset management) designed to drive the development, build, and release processes thanks to a set of jBPM workflow processes, which kick in at various stages. Managed repositories, in short, provide project Maven version control and Git branch management.

Governance workflow

The asset management workflows are not fully automatic; they also require a managing actor (who must be in the `kiemgmt` role; see the KIE workbench `roles.properties` file) to complete specific tasks in order to make the workflow progress (selecting assets to release and/or review) or present the user with informative data (such as error data). The workflows trigger only when the following pre-defined operations take place (remember, this only applies to managed repositories):

- **Repository creation or configuration**: After the Git repository is created or when the repository **Configure** button is selected, the workflow can automatically add a `dev` branch and a `release` branch for you (the `master` branch is always the default one).

- **Asset promotion**: When a user thinks that his/her assets are ready to be released, he/she can submit them for a so-called **promotion** by selecting the **Promote** button. The promotion requires the managing user to select (Git cherry picking) and approve the submitted changes by promoting them to the Git release branch, or to delay the process for a later review.

- **Project build**: The user performs **Build**, selecting a specific repository branch. The build involves the compiling and installing of the project as a Maven module into the internal Kie repository.

- **Release**: The user performs **Release**. The release feature involves the building and deploying processes at the repository level. All of your repository projects are built and then, published to the Kie runtime (the **Deploy To Runtime** option) so that business assets can be used. We can only release from branches starting with the **release** label (for example, release-1.0.1).

> Please check Chapter 9 of the jBPM 6.2 User Guide for additional details on asset management workflows.

Git cloning and managed repository

jBPM 6.2 does not support importing into a managed repository; the naïve solution mimics what we have seen in the (unmanaged) *New repository* paragraph with slight variations; let us see how to import our GitHub `chapter4-managed` repository projects:

Let us create a multi-module repository; the settings for each step are as follows:

1. **Repository Name**: `ManagedVesuvio`, and **Organizational Unit**: `demo` (not relevant now).

2. **Multi-project Repository** (checked), **Automatically Configure Branches** (checked), and **Project Settings** (leave defaults).

> In Step 2, the specified Maven GAV will be the Maven parent module GAV (demo:ManagedVesuvio:1.0.0-SNAPSHOT). Two additional branches are created: dev-1.0.0 and release-1.0.0. If we were not selecting automatic branch management, only the default master branch would be available and the repository could not be released as a whole (see the *Releasing* section for additional hints).

3. Clone the chapter4-managed example projects:

    ```
    git clone https://github.com/masteringjbpm6/chapter4-managed.git
    ```

4. Add a new remote named kievesuvio (or if you prefer, replace the origin):

    ```
    git remote add kievesuvio ssh://admin@localhost:8001/
    ManagedVesuvio
    ```

5. Change into the chapter4-managed folder, and add the files and commit:

    ```
    git add .
    git commit -m "first Kie managed commit"
    ```

6. Get updates from the master branch and push to the KIE ManagedVesuvio dev-1.0.0 branch:

    ```
    git pull kievesuvio master
    git push kievesuvio master:dev-1.0.0
    ```

*   At this stage, the dev-1.0.0 branch is updated, but our ManagedVesuvio repository structure is not aware of the branch changes while the KIE Project **Explorer** is. This is due to the fact that the repository structure shows the Maven multi-module configuration (more details in the *Managed repository and Maven* section) and that its pom.xml file is stale. We have to add the napoli and vesuvio projects manually to it.

7. In File Explorer (**Authoring | Administration**), click on pom.xml and add the <modules> element as follows:

    ```
    <modules>
      <module>napoli</module>
      <module>vesuvio</module>
    </modules>
    ```

After saving the file, KIE should pick up the projects and the repository structure should display our modules.

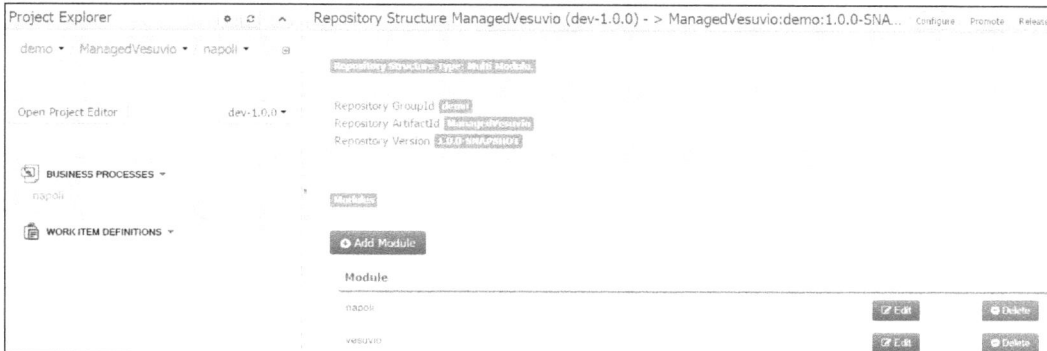

The repository is now properly configured, and we are ready to tackle the asset management features.

Asset management example

The `napoli` and `vesuvio` projects each contain a basic process definition. The `napoli` process (`napoli.bpmn2`) includes the `vesuvio` process (`vesuvio.bpmn2`) as a reusable sub-process (more on BPMN2 elements in *Chapter 5, BPMN Constructs*). The users perform asset management tasks by selecting the appropriate button in the **Repository | Repository Structure** view (see the following screenshot) and by completing human tasks in the **Task | Task List** window.

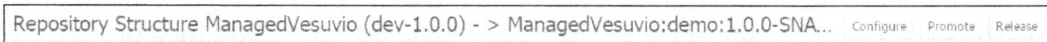

The asset management tasks are assigned to users owning the `kiemgmt` role only; therefore, ensure that you log into the KIE workbench with the `admin` user since this is the only preset user owning this role.

Promoting assets

We submit the assets (the napoli and vesuvio modules) for promotion to the release branch:

1. Select **Project | Authoring and Repository | Repository Structure**; select the **Promote** button and enter the target branch: `release-1.0.0`.

2. In the **Tasks | Task List** window, you should now be assigned a **Select Assets to Promote** task; click on it, **Claim** the task, **Promote All** assets, and **Complete** the task as shown in the following screenshot:

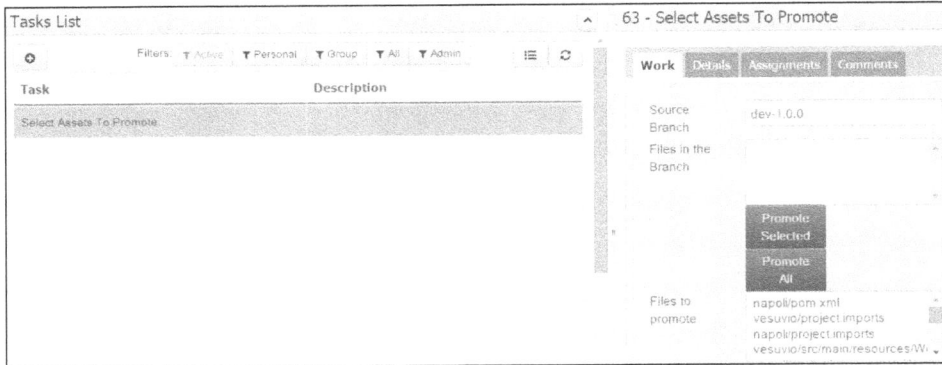

3. Head back to **Repository Structure**, and by selecting the release-1.0.0 branch, you should see the two promoted projects.

> Assets are now merged into the Git repository release-1.0.0 branch.

Releasing

In **Repository Structure**, select the release-1.0.0 branch and press the **Release** button. Optionally, bump the Release Version for the module, toggle **Deploy to Runtime** (user: admin, password: admin, Server URL: default), and then proceed.

> napoli and vesuvio are now installed into the KIE Maven repository, and you can find them among your **Authoring | Artifact Repository** artifacts. In case you selected **Deploy to Runtime**, the contained process definitions would be made available in **Process Management | Process Definitions**.

Building (single project)

The release process always runs through a build process for all the managed repository projects; the build/release process can also be performed, for all kind of repositories (managed/unmanaged), on the single project by the **Build** menu:

- **Build & Install**: Deploy the artifact to the Kie repository and the system Maven repository (if any)

- **Build & Deploy**: Perform the install step (see previously), and then, deploy the module to the Kie runtime: business artifacts are available for runtime usage
- We will discuss more on deployments in the *KIE deployments* section

Asset versioning history

All Git versioning information for the assets is available in the **Overview** tab. For instance, by clicking on the **napoli** process, we can see all the commit logs and we can load the previous Git versions for the assets with the **Select** and **Current** buttons (see the following screenshot; the comments may vary).

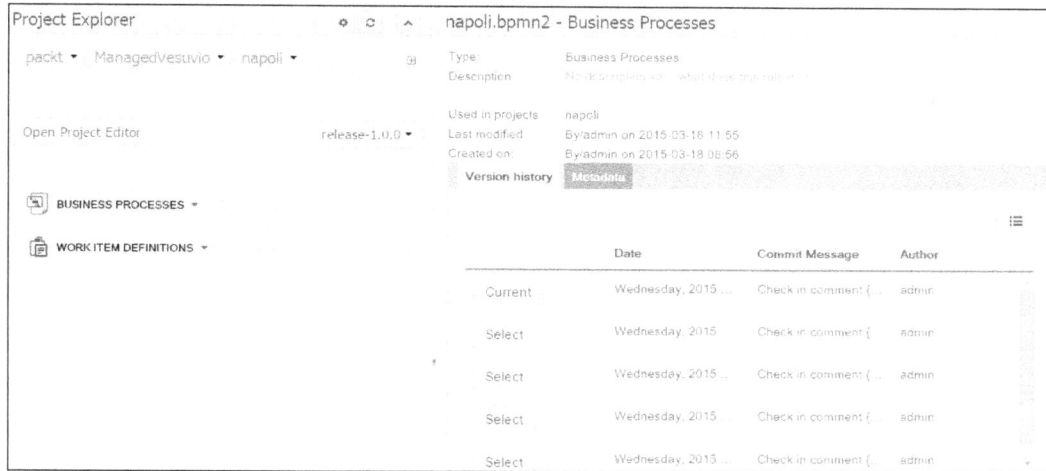

Managed repository and Maven

The managed Git repositories are themselves stored as Maven modules in the `repositories/kie` repository. They can be Single Project or Multi-Project repositories; this affects the way that the Mavenized Kie projects are configured.

- The Single Project repository contains a single Kie Maven project
- The Multi-Project repository contains a Maven multi-module project

The parent module `pom.xml` file shows that it is a `pom` (multi) module containing the `napoli` and `vesuvio` modules:

```
<groupId>packt</groupId>
<artifactId>ManagedVesuvio</artifactId>
<version>1.0.0-SNAPSHOT</version>
<packaging>pom</packaging>
```

```
<name>ManagedVesuvio</name>
<modules>
  <module>napoli</module>
  <module>vesuvio</module>
</modules>
```

The `napoli` module `pom.xml` file shows that it depends on the `vesuvio` module:

```
<parent>
  <groupId>packt</groupId>
  <artifactId>ManagedVesuvio</artifactId>
  <version>1.0.0-SNAPSHOT</version>
</parent>
<groupId>packt</groupId>
<artifactId>napoli</artifactId>
<version>1.0.0-SNAPSHOT</version>
<name>napoli</name>
<dependencies>
  <dependency>
    <groupId>packt</groupId>
    <artifactId>vesuvio</artifactId>
    <version>1.0.0-SNAPSHOT</version>
  </dependency>
</dependencies>
```

> Please refer to `http://maven.apache.org/guides/mini/guide-multiple-modules.html` for an introduction to Maven multi-module management.

Governance process definitions

The jBPM asset management process definitions and their related business logic can be found in the Drools project called `guvnor-asset-mgmt`. This module is pre-deployed and is loaded by the KIE workbench; it is regularly listed in **Artifact Repository**, and you can find its process definitions in the **Deployments | Process Deployments** window and their running instances in the **Process Instances** window.

A final note

The KIE workbench project editor only lets you edit, build, and release the proper Kie modules and not plain Maven modules (which lack the `kmodule.xml` file). So, if you only have a single Kie module, it does not make much sense to clone a bunch of modules into a Kie multi-project managed repository; try to design your repository content with a grain of salt so that your Kie runtime and development environment is always clean and healthy. Now, instead of using the KIE workbench to complete the configuration for our next example (left for an exercise to the reader), we switch to an alternative tool, the KIE CLI (which stands for command line interface), and see different ways to perform the common operation tasks.

An overview of the kie-config-cli tool

Chapter 2, *Building Your First BPM Application* (see the *Creating your first jBPM project* section) introduced several KIE console features (repositories and organizations management, module deployment, and so on). The KIE CLI utility provides the administrator/user with the tools to perform common tasks related to organizational units, repository, user permissions, and deployment management at the console/shell level.

> The tool project is hosted at GitHub: `https://github.com/droolsjbpm/kie-wb-distributions/tree/master/kie-config-cli`.
>
> You can download the tool distributions from the JBoss snapshot Maven repository:
>
> `http://snapshots.jboss.org/maven2/org/kie/kie-config-cli`.

The previous section explained in detail how KIE handles Git repositories for source projects, and we previewed the existence of a KIE system Git repository; the KIE CLI tool interfaces with it and operates in two modes:

- **online (default and recommended)**: Upon startup, it connects to the system repository by using the Git service embedded in `kie-wb`. All changes are local and published to the upstream only when the `push-changes` CLI command is explicitly executed. The `exit` command will publish all local changes; to discard local changes upon exiting the CLI, the `discard` command shall be used

- **offline**: Creates and manipulates the Kie system repository directly on the server (no discard option is available)

The system repository stores the private configuration/settings data for the KIE workbench: how editors behave, organizational groups, security, and so on.

The system repository is located in the KIE `.niogit` folder (`.niogit/system.git`).

> The core Git backend features are provided by the Red Hat Uberfire framework. `http://www.uberfireframework.org`.

By default, the KIE workbench monitors the `system.git` repository changes, thanks to its backend services, and updates its UI accordingly. We will continue our system repository description in the next section after our next example environment has been set up. We could have accomplished the job through the KIE workbench features, but we want you to go hands on and put the KIE console at work, getting acquainted in what is going on behind the curtains.

Connecting (online mode)

After launching the tool (by the `kie-config-cli` script in the installation folder), let us start by connecting to the KIE system repository on the localhost.

> SSH is to be preferred over the Git protocol for security reasons.

```
************* Welcome to Kie config CLI ****************
>>Please specify location of remote Git system repository [ssh://
localhost:8001/system]
ssh://localhost:8001/system
>>Please enter username:
admin
>>Please enter password:
admin
```

Creating an organizational unit

The organizational unit is required by the KIE workbench in order to create repositories and users and have RBAC rule control that is entitled to perform certain tasks.

```
create-org-unit
>>Organizational Unit name:packt
```

```
>>Organizational Unit owner:admin@packt.org
>>Default Group Id for this Organizational Unit:com.packt
>>Repositories (comma separated list):
Result:
Organizational Unit packt successfully created
```

Creating a repository

We have already seen how to clone a Git repository from the KIE workbench. Let us now create a new Git repository (local, bare, and unmanaged; no username/password required) in which we can store our new projects.

```
create-repo
>>Repository alias:masteringjbpm6
>>User:
>>Password:
>>Remote origin:
Result:
Repository with alias masterjbpm6 has been successfully created
```

Defining role-based access control rules

To complete the setup procedure for our example, we set some RBAC rules to our masterjbm6 repository; the organization has no roles set, so in order to constrain the access to the repository, we add roles to the repository object:

```
add-role-repo
>>Repository alias:masteringjbpm6
>>Security roles (comma separated list):user, analyst
Result:
Role user added successfully to repository masterjbpm6
Role analyst added successfully to repository masterjbpm6
```

Adding jBPM console users

`users.properties` and `roles.properties` define the users (with their roles) that are enabled to log into the jBPM console.

Let us edit `users.properties` and add two new users:

```
simone=simone
arun=arun
```

Edit `roles.properties` and associate the users with the roles that we have created in the previous step:

```
simone=admin
arun=admin
```

Adding the repository to an organization

Each KIE Git repository must be bound to an organization; the organization is in charge of controlling accesses to it and giving organization users operation permissions.

add-repo-org-unit

>>Organizational Unit name:packt

>>Repository alias:masteringjbpm6

Result:

Repository masteringjbpm6 was successfully added to Organizational Unit packt

Pushing changes to system.git

The `push-changes` command sends changes to the KIE `system.git` repository:

push-changes

>>Result:

>>Pushed successfully

As a consequence, your KIE workbench gets a refresh, displaying the updated settings on your UI.

More on the system.git repository

At this point, you should have an idea of what the `system.git` repository is for; as a final exercise, let us clone it and have a look inside it:

```
git clone file:///$JBPM_HOME/.niogit/system.git
```

> Remember to never push changes to system repository from outside the KIE CLI tool; it would likely mess up the entire KIE workbench installation!

The system repository contains some entities (organizational units and repositories) and internal configuration files; here, we can find our brand new organization and repository description files:

- `masteringjbpm6.repository`
- `packt.organizationalunit`

The `masteringjbpm6.repository` file content is as follows (please note the `security:role` settings):

```
<group>
  <name>masteringjbpm6</name>
  <description></description>
  <type>REPOSITORY</type>
  <enabled>true</enabled>
  <items>
    <entry>
      <string>scheme</string>
      <item>
        <name>scheme</name>
        <value class="string">Git</value>
      </item>
    </entry>
    <entry>
      <string>security:roles</string>
      <item>
        <name>security:roles</name>
        <value class="list">
          <string>user</string>
          <string>analyst</string>
        </value>
      </item>
    </entry>
    <entry>
      <string>branch</string>
      <item>
        <name>branch</name>
        <value class="string">master</value>
      </item>
    </entry>
  </items>
</group>
```

Now that we are done with our new KIE Git repositories and environment configuration, we are ready to tackle the new deployment feature and the Kie-Maven integration, the subject of our next example.

KIE deployments

The jBPM 6 platform introduced a brand new deployment process; the previous proprietary mechanism that leveraged the Guvnor packages (backed by a **Java Content Repository (JCR)** and the Drools KnowledgeAgent (changeset.xml) was replaced with the widely adopted Apache Maven tool. This greatly improved the development process both in terms of tool configuration (more convention/configuration oriented) and support, standardization, and deployment flexibility.

When you deploy your project, you physically create a KIE deployment unit (KJAR); this module is a Maven-enabled project and is a compressed standard Java archive that contains all the project's business assets (processes, workitem handlers, business rules, forms, and so on) as well as its knowledge session and runtime declarative metadata descriptor (META-INF/kmodule.xml).

> The kmodule.xml file is extensively covered in the official jBPM and Drools documentation.

The unique ID of a KIE module is built starting from its Maven GAV (GroupId, ArtifactId, Version) with the addition of the knowledge base name (the default knowledge base name is empty; we will return to this in *Chapter 6, Core Architecture*), for example:

```
groupID:artifactID:version{:kbasename}
```

The jBPM runtime resolves KJAR dependencies while automatically searching for other Maven modules in the configured Maven repositories (either by the project pom.xml file embedded in the KIE JAR or through the Maven settings.xml file) using the Drools KIE-CI components. You can alternatively use the kie.maven.settings.custom system property and point to any Maven settings.xml file.

> Refer to *Chapter 6, Core Architecture*, particularly to the *Repositories and scanner* section, for detailed information on class loading and resolving module dependencies at runtime.

The default KIE Maven repository artifacts (the repositories/kie folder) are remotely accessible at the following URL: http://{jbpmconsole-host}:{port}/jbpm-console/maven2wb/.

Let us now summarize the core actions that are performed during the deployment process:

- Maven install of the module into the KIE Maven repository

- Maven deploy of the module into your system Maven repository (the Maven `settings.xml` file from the Maven home is used, or the `kie.maven.settings.custom` system property is checked).

- The jBPM database table called `DeploymentStore` is updated with the deployment descriptor (in the XML format). This change has been introduced with the jBPM 6.2 release; prior to this, the deployment information was stored inside the `system.git` repository.

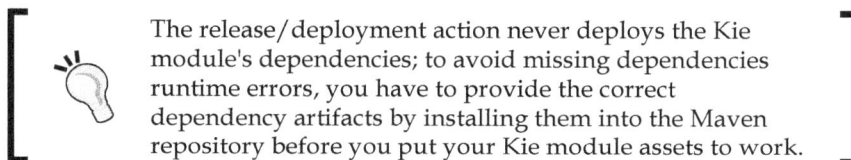

> The release/deployment action never deploys the Kie module's dependencies; to avoid missing dependencies runtime errors, you have to provide the correct dependency artifacts by installing them into the Maven repository before you put your Kie module assets to work.

The following figure captures the standard Maven repository configuration when working with the KIE console; the KIE project is deployed into the internal KIE Maven repository first, which synchronizes with the remote **Maven Repository**, and then, it provides remote public access (HTTP) to any application in order to resolve its dependencies.

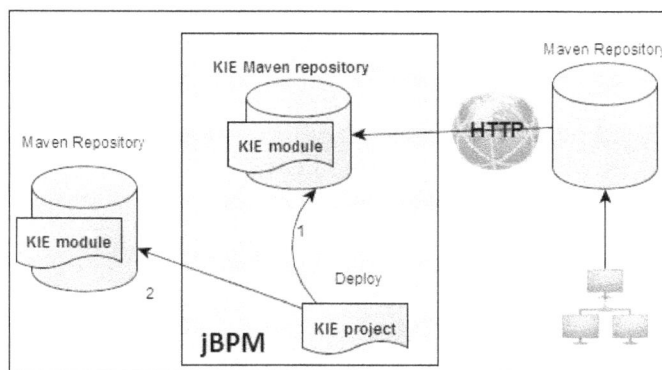

Artifacts and KIE modules

A KIE module may depend on a number of additional Maven artifacts. The KIE workbench provides you with a view of your KIE Maven repository by selecting the **Authoring | Artifact Repository** menu item. The **Path** column displays the Maven repository URL for the artifact; just to clarify, let us have a look at the napoli Kie module that we released from our ManagedVesuvio repository:

Artifact Name: napoli-1.0.0.jar

Path: packt/napoli/1.0.0-SNAPSHOT/napoli-1.0.0.jar

This artifact is available at the following Maven artifact URL:

```
http://{jbpmconsole-host}:{port}/jbpm-console/maven2wb/packt/
napoli/1.0.0-SNAPSHOT/napoli-1.0.0.jar
```

Remember that while **Artifact Repository** is a snapshot of the KIE Maven repository content, the KIE **Deploy | Deployments** menu item exclusively displays the valid KIE modules (KJARs), which are loaded and validated from the KIE console runtime.

> org.guvnor.m2repo.dir: System properties set the path where the Maven repository folder will be stored; the default is ${jBPM-install-directory}/repositories/kie.

Deployment options

The KJAR artifacts must always be deployed to the KIE console's Maven repository so that we can centrally manage them and have the console behave consistently. Given this, thanks to both the introduction of the Maven repository style and the new breed of tools (the KIE console and the Eclipse BPMN tools), you can tailor the deployment process to your development environment with a nice degree of flexibility.

Here you have some viable deployment options:

- Create a project (Mavenized by default) from the KIE console into a configured KIE Git repository; Git clone and pull from Eclipse, continue the development from Eclipse (add business models and so on), push changes to KIE (the KIE console automatically refreshes its repository view); build and deploy from KIE.

- Create a Maven project from Eclipse (remember to add the `kmodule.xml` file and the jBPM dependencies in `pom.xml`); create all your business artifacts, and develop a unit test from Eclipse; push to KIE; maybe make some fixes from KIE and then build and deploy.

- Create a Maven project from Eclipse as in the previous solution; install with Maven (either from Eclipse or from the command line); have the KIE console advertise the new KIE deployment module uploading the new artifact from the console (**Deploy** | **Deployments**).

- From the Kie console, create a repository clone from a remote Git repository; create a project, add assets, and then, save and commit. Deploy to Maven and push changes to the remote Git repository.

- Thanks to its Maven and Git integration, the KIE platform can fit very flexibly into a complex development environment.

Deployment by example – the Pizza projects

We had a preview of installing and deploying with the `ManagedVesuvio` repository release process where two KIE modules (one is dependent on the other) were released. Let us go hands on with a different module example: a KIE module (main project) and a plain module (dependency). These pizza example projects are as follows:

- `Pizzadelivery`: The KJAR module (it contains the process definition)
- `Pizzamodel`: A utility project with Java classes that model our business objects (Order, Pizza, and so on)

The `Pizzadelivery` project depends upon the `Pizzamodel` project. Let us start by reviewing the example process definition (we will use the same process for our BAM example solution later, in the closing paragraph).

The process definition – pizzadelivery

The example process definition captures a typical takeaway pizza process:

1. An order is placed and Nino manages the incoming order by phone (the **Handle Incoming Order** task).
2. Maria gets the order details from Nino and hands off a sticky note to the pizza makers (the **Order Assignment** task).

3. A pizza maker (either Mario or Luigi) starts preparing the pizza (the **Make the Pizza** task), while, at the same time, Maria assigns the order delivery to a pizza boy (the **Assign the Delivery** task). The completion of both the delivery assignment task and the making pizza task (parallel tasks, more on this in *Chapter 5, BPMN Constructs*) means that the pizza is ready to be delivered.

4. Salvatore delivers the pizza (the **Pizza Delivery** task).

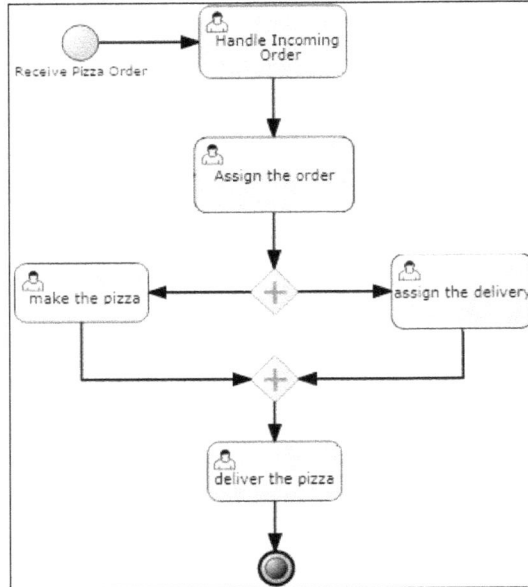

Deploying pizzas

First, we make the dependency available through Maven, and then, we deploy the module through the KIE workbench. Let us import the sources into our new KIE repository (the one we set up in the *Creating a repository* section).

Git and project configuration

At this stage, you should be quite familiar with importing a source project into the KIE repository:

1. Clone the examples repository from our example GitHub repository:

```
git clone https://github.com/masteringjbpm6/chapter4-pizza.git
```

2. Add the cloned repository projects to the local (unmanaged) `masteringjbpm6` repository section:

```
git remote remove origin
git remote add origin ssh://localhost:8001/masteringjbpm6/
```

3. By issuing a `git remote` command, we see the following:

```
$ git remote -v
origin ssh://admin@localhost:8001/masteringjbpm6 (fetch)
origin ssh://admin@localhost:8001/masteringjbpm6 (push)
```

4. Let's now push only the sample KIE module project to the new remote (origin):

```
git add pizzadelivery
git commit -m "pizzadelivery: first kjar"
git push -u origin master
```

The aim here is to send the KIE console (through Git) only the KIE project, and not to create additional sources of issues. We are now going to provide the KIE `pizzadelivery` project Maven dependencies (`pizzamodel.jar`) through our Maven repository.

Deploying the dependency

By switching to **Project Authoring** | **Project Explorer**, we can find the PizzaDelivery project (navigate through the **packt/masteringjbpm6** repository breadcrumb). Press the **Open Project Editor** button, and by selecting the **Project Settings** | **Dependencies** item from the dropdown list, we see that the `pizzadelivery` module depends on the `pizzamodel` artifact (see the picture below), which is not present in the Maven repository yet.

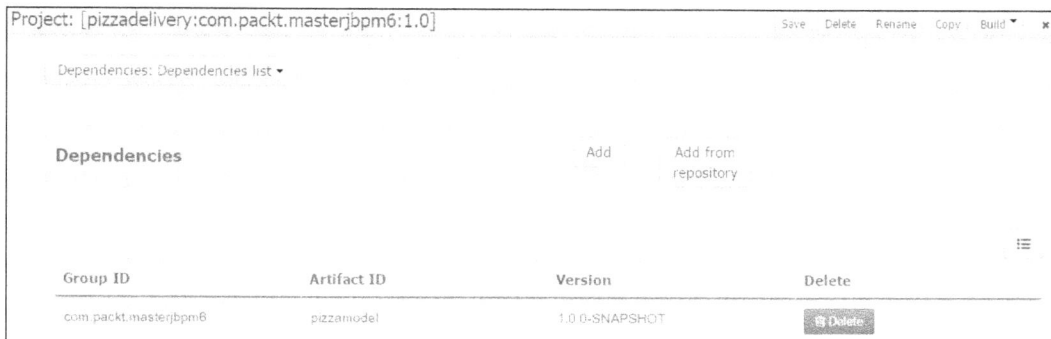

The **Messages** tab reports this issue, accordingly:

Unresolved dependency com.packt.masterjbpm6:pizzamodel:1.0.0-SNAPSHOT

Providing artifacts

What is going on now should be clear: the KIE dependency mechanism (the "scanner" we will talk about it in *Chapter 6, Core Architecture*) cannot resolve the `pizzadelivery` project's dependencies (loaded from its `pom.xml` file) while scanning through the available Maven repositories; to address the issue, we are going to supply the missing artifact with two steps:

1. Maven install: Executing an `mvn clean install` at the `pizzamodel` project root (or using the Eclipse **Run As Maven Build** integrated feature) performs the build and install of the artifact in our Maven repository so that the dependency can be resolved at runtime.

2. Artifact repository upload: Go to the Kie console **Authoring | Artifact Repository** page and click the **Upload** button; select the `pizzamodel` JAR file from your Maven repository folder (`com/packt/masterjbpm6`). The KIE console will copy the artifact to its internal Maven repository.

Name	Path	LastModified	Open	Download
guvnor-asset-mgmt-project-6.2.0.Fi...	org/guvnor/guvnor-asset-mgmt-project/6.2.0.Final/guvnor-asset-mgmt-project-6.2.0.Final.jar	2015 Mar 22 18:26:37	Open	Download
pizzamodel-1.0.0-20150322.17085...	com/packt/masterjbpm6/pizzamodel/1.0.0-SNAPSHOT/pizzamodel-1.0.0-20150322.170854-...	2015 Mar 22 18:08:54	Open	Download

The artifact is now available, and if you hit the **Refresh** button in the **Messages** panel, the issue should be fixed.

Deploying

We are now ready to deploy. In the **Projects Explorer** tab, select the **Tools | Project Editor** menu item. Check whether the `PizzaDelivery` module Maven GAV properties are correct and click the **Build & Deploy** button.

Please note that during deployment, the KIE console will try to resolve and verify all of your project dependencies, potentially hitting a number of remote Maven repositories (depending on your configuration) and taking some time to finish. Ensure that your Internet connection is active or set up a Maven proxy repository (highly recommended).

The application server console traces the following information:

```
18:32:33,517 INFO   [org.drools.compiler.kie.builder.impl.
KieRepositoryImpl] (default task-106) KieModule was added:
MemoryKieModule[releaseId=com.packt.masterjbpm6:pizzadelivery:1.0]

18:32:34,547 INFO   [org.jbpm.console.ng.bd.backend.server.
DeploymentManagerEntryPointImpl] (default task-106) Deploying unit com.
packt.masterjbpm6:pizzadelivery:1.0

18:32:35,969 INFO   [org.jbpm.kie.services.impl.store.
DeploymentSynchronizer] (default task-106) Deployment unit com.packt.
masterjbpm6:pizzadelivery:1.0 stored successfully
```

> Remember that you cannot deploy a deployment unit having the same ID (overwriting it), irrespective of whether it has active (running) process instances or not; an undeploy action is required (see the next section).

The logs confirm that the deployment was successfully completed; we can see our KJAR module listed in the **Deploy** | **Deployments** tab. Remember that for the deployment process to succeed (at least with jBPM releases up to 6.2), your `kmodule.xml` must either:

- Declare an empty `<kmodule>` element
- Declare a `<ksession>` element with the following attributes:
  `type="stateful" default="true"`

Deployed Units				× ▾
+				≣ ⟳
Deployment	Runtime strategy	Status	Actions	
com.packt.masterjbpm6.pizzadelivery.1.0	SINGLETON	Active	⊚ ⊘	

For the sake of thoroughness, check your jBPM data store for the deployment entry in the DEPLOYMENTSTORE table; we should see that a new row has been added.

The table row has a DEPLOYMENTID column: `com.packt.masterjbpm6:pizzadelivery:1.0.0` and a DEPLOYMENTUNIT column, which contains the actual deployment descriptor:

```
<org.jbpm.kie.services.impl.KModuleDeploymentUnit>
  <artifactId>pizzadelivery</artifactId>
  <groupId>com.packt.masterjbpm6</groupId>
  <version>1.0.0</version>
  <strategy>SINGLETON</strategy>

</org.jbpm.kie.services.impl.KModuleDeploymentUnit>
```

This is, actually, the third step of the deployment process (see the *KIE deployments* section).

Adding a KIE module manually

Even though you have built and installed your KIE module from outside the KIE workbench (to your system Maven repository only, maybe using Eclipse IDE and a Maven install goal), you can always deploy it to KIE later. Manually adding a deployment unit means that you are making this (Maven installed) module available to the KIE runtime. From the **Deploy** | **Deployments** perspective, you can add new deployment units (the **New Deployment** button); just provide the Maven GAV for the project you want to deploy and, optionally, the knowledge base and the knowledge session name.

> In addition to this, you can select the `ksession` runtime strategy that fits your requirements: a Singleton, Per Request, or Per Process instance (See *Chapter 6, Core Architecture* for runtime strategies).

The KIE workbench will provide a new `kmodule.xml` file and meta information, converting your plain JAR module to a new KIE module. You cannot create a new deployment unit if the matching Maven artifact is not available in the KIE Maven repository.

Process and task management applied – the PizzaDelivery process

The KIE workbench allows you to manage process instances and interact with process tasks. You can do the following:

- Send a signal to a specific process instance or a bulk signal (broadcast to all process instances).
- Abort a specific process instance or perform a bulk abort – the abort terminates the instance(s) and all the pending tasks. All data pertaining to the process and its tasks are removed from the jBPM database tables.
- Get the process details – the details page includes the auditing log taken from the default jBPM database auditing tables (we will talk about auditing in the last section and in *Chapter 6, Core Architecture*).
- Start, release, and complete a task.

Starting the Process

Open the **Process Management | Process Definitions** tab and click the **Start** icon button next to our brand new deployed PizzaDelivery process definition; a new process instance starts.

Undeployment

The undeployment action removes the deployment unit's configuration file only from the jBPM database table, leaving the Maven artifact in place.

User management

We introduced the PizzaDelivery process in the example setup section; the process requires five different human actors:

- **Incoming orders management**: Nino
- **Order assignments**: Maria
- **Pizza cooking**: Either Mario or Luigi
- **Delivery assignments**: Maria
- **Pizza deliveries**: Salvatore

In order to use the KIE console to exercise our first process instance, we need to add these process participants (the actors) to the KIE runtime. Edit the user.properties and the roles.properties files in the $JBOSS_HOME\standalone\configuration folder; these users will be automatically added by the KIE workbench to the task service database table (ORGANIZATIONALENTITY, more on this in *Chapter 6, Core Architecture*).

Add the actors (specify the authentication password) to the user.properties file:

```
nino=nino
maria=maria
salvatore=salvatore
mario=mario
luigi=luigi
```

Specify a role for the users in the roles.properties file (the default role user is enough to perform tasks):

```
nino=user
salvatore=user
mario=user
luigi=user
maria=user
```

> There is no need to restart the application server to make it pick up the new settings.

Task management

To complete the process instance, perform the steps in the following order:

- **Log in as Nino**: **start** and **complete** the **Handle Incoming Order** task
- **Log in as Maria**: **claim** and **complete** the **Assign Order** task
- **Log in as Mario (or Luigi)**: **claim** and **complete** the **Make Pizza** task
- **Log in as Maria**: **claim** and **complete** the **Assign Delivery** task
- **Log in as Salvatore**: **claim** and **complete** the **Deliver the pizza** task

All tasks are completed. The process instance ends and looking at the process instance detail tab, we can see the instance log traces (events are displayed in the reverse chronological order):

Instance log:

```
22/nov/14 23:35:53: 8 - EndNode
22/nov/14 23:29:54: 7 - deliver the pizza (HumanTaskNode)
22/nov/14 23:29:54: 6 - Join
22/nov/14 23:27:50: 6 - Join
22/nov/14 23:26:56: 4 - make the pizza (HumanTaskNode)
22/nov/14 23:26:56: 5 - assign the delivery (HumanTaskNode)
22/nov/14 23:26:56: 3 - Split
22/nov/14 22:41:56: 2 - Assign the order (HumanTaskNode)
22/nov/14 18:10:05: 1 - Handle Incoming Order (HumanTaskNode)
22/nov/14 18:10:05: 0 - Receive Pizza Order (StartNode)
```

The `Split` log traces the activation of the parallel diverging gateway. The `Join` logs trace the activation of the parallel converging gateway's incoming connections. The trace logs are loaded from the `NODEINSTANCELOG` table.

As you may have realized, it is not that easy to perform a full test of a process definition from within the KIE workbench; switching from actors back and forth is a cumbersome and time-consuming task…just think about a complex process with a lot of human tasks and actors or groups. We will see how to overcome these issues by using test automation with the BAM example and in the next chapter.

Managing jobs and asynchronous commands' execution

Starting from jBPM 6, the platform features a new scheduler service (called `Executor`), which lets you schedule, execute, and manage asynchronous jobs tasks. Executor can be used either as a general-purpose Java batch scheduling facility or as a service able to execute asynchronous process tasks (see *Chapter 5, BPMN constructs, the Async task* section for more details). The asset management feature, for instance, internally schedules different types of commands (to get an idea, open the **Deploy | Jobs** window, as shown in the following screenshot): `CreateBranchCommand`, `ListCommitsCommand`, `BuildProjectCommand`, `MavenDeployProjectCommand`, and so on.

Requests List				
Actions ▾	Showing ▼All ▼Queued ▼Running ▼Retrying ▼Error ▼Completed ▼Cancelled			
Id	Type	Status	Due On	Actions
1	org.guvnor.asset.management.backend.command.CreateBranchCommand	DONE	Thu Mar 12 16:18:46 GMT+100 2015	Details
2	org.guvnor.asset.management.backend.command.CreateBranchCommand	DONE	Thu Mar 12 16:18:49 GMT+100 2015	Details
3	org.guvnor.asset.management.backend.command.CreateBranchCommand	DONE	Fri Mar 13 09:36:51 GMT+100 2015	Details
4	org.guvnor.asset.management.backend.command.CreateBranchCommand	DONE	Fri Mar 13 09:36:54 GMT+100 2015	Details
5	org.guvnor.asset.management.backend.command.ListCommitsCommand	DONE	Fri Mar 13 14:07:03 GMT+100 2015	Details
6	org.guvnor.asset.management.backend.command.BuildProjectCommand	DONE	Fri Mar 13 14:16:54 GMT+100 2015	Details
7	org.guvnor.asset.management.backend.command.MavenDeployProjectCommand	DONE	Fri Mar 13 14:16:57 GMT+100 2015	Details

The Executor service executes preconfigured `Command` classes; a `Command` is a Java class that executes a set of business statements running outside the jBPM process context and communicating with Executor through a set of interfaces (`CommandContext` and `ExecutionResults`), which enforce parameter passing.

The `Job` classes are persisted in the `REQUESTINFO` jBPM database table, while the resulting errors problems are persisted in the `ERRORINFO` table.

Creating, scheduling, and launching a new Job

The plain and simple, general-purpose `Job` definition (no jBPM context available) requires you to provide at least the class name for the class to schedule (see the following screenshot).

1. Type the class name com.packt.masterjbpm6.command.SimpleCommand into the **Type** field and SimpleCommand in the **Name** field. The Job class must be in the classpath for the KIE workbench application (jbpm-console.war), so either copy the pizzamodel.jar file in WEB-INF/lib for the exploded WAR or copy it in the dependencies folder of the jBPM setup folder and rebuild the console app by using the Ant target install.jBPM-console.into.jboss you can find in the jBPM build.xml file.

2. Set the **Due On** (schedule) time and, optionally, the number of **Retries** (the number of times the Job class can be restarted after failure), and the parameters.

3. The parameters (the contextual data) are passed upon execution to the Job instance through the CommandContext class. Parameters must be serializable.

Quick New Job

* Name

SimpleCommand

Due On

15/03/2015   10:43

* Type

com.packt.masterjbpm6.comman

* Retries

0

Key	Value	Actions
click to edit	click to edit	Remove
Add Parameter		

Create

After being created, the task turns into the **QUEUED** state (as shown in the following screenshot) and will be executed at the scheduled time. The different Job statuses are as follows: **QUEUED, DONE, CANCELLED, ERROR, RETRYING**, and **RUNNING**.

Requests List   Settings   New Job   Refresh   ✕   ▾

Showing   All   Queued   Running   Retrying   Completed   Cancelled   Error

Id	Job name	Status	Due On	Actions
33	com.packt.masteringjbpm6.pizzadelivery.command.SimpleCommand	QUEUED	Tue Nov 11 09:57:00 GMT+100 2014	Details Cancel

Process definition conversion

Conversion deals with moving a process definition from an old version format to a new version format. jBPM 6 gives us some (pretty much experimental) options when upgrading older process definitions to a new jBPM release format:

- Importing from the old proprietary jBPM JPDL 3/4 to BPMN2 with the jBPM web process designer menu function

- Ad hoc migration with `jBPM5migrationtoolproject` and `jbpmmigration-0.13.jar` or a newer release (API mapping is also supported)

The goal of the `jBPM5migrationtoolproject` project is to provide some migration tooling to the existing users of jBPM for moving from jBPM5.

> The project home and Wiki pages are available here: `https://developer.jboss.org/wiki/jBPM5migrationtoolproject`. The project is hosted on GitHub: `https://github.com/droolsjbpm/jbpmmigration`.

Process definition versioning and instance upgrading

Depending on enterprise business requirements and business organizations, the processes may change at a very variable rate over time; several business migration cases need to be addressed:

- A complex critical business process may take months to complete (maybe due to manual tasks), yet the business staff needs to make an updated process definition available as soon as possible because some old legacy systems must be integrated into the flow

- The process definition needs a fix but a number of instances of that very same process definition are active and we do not want to abort them and have the user restart the workflow from the beginning

Apart from bumping the process definition version property (numeric), which is just a mnemonic and does not affect the process instance behavior, it's good practice to name your process ID (string) in order to reflect the version number, since the engine itself does not provide any version tracking mechanism, for example:

```
com.packt.masteringjbpm6.pizzadelivery_v1_0,
com.packt.masteringjbpm6.pizzadelivery_v1_1
```

This way you can have a flexible method of switching instantiations across different versions of the process definition while preserving auditing data and maintaining separation. Each instance is, in fact, bound to its process definition (by the ID), and this must not be overwritten until the instance is completed.

To support users in migrating a process instance across different process definitions, jBPM 6 features the WorkflowProcessInstanceUpgrader class.

The WorkflowProcessInstanceUpgrader.upgradeProcessInstance method first disconnects the process instance from the signals, and event handler management then traverses the process instance node-by-node trying to map the nodes to the target process definition nodes by getting uniqueID from the mapping data that you provide.

uniqueID is an internal identifier generated by the engine by concatenating the IDs of the parent (container) elements of the node, for example:

```
// create the node mapping data
Map<String, Long> mapping = new HashMap<String, Long>();
// top level node 1 must be mapped to a new node with id 2
mapping.put("1", 2L);
// node 3, which is inside a composite node 4, must be mapped to a new
node with id 5
mapping.put("4.3", 5L);
// upgrade old processInstance to a new process instance definition
with id= com.packt.masteringjbpm6.pizzadelivery_v1_1
WorkflowProcessInstanceUpgrader.upgradeProcessInstance( ksession,
processInstance.getId(),"com.packt.masteringjbpm6.pizzadelivery_v1_1",
mapping);
```

This solution is, all in all, far from complete for complex process definitions; you are suggested to implement your own process migration whenever possible.

BAM

The business activity monitor (BAM) provides the tools to build out of the system, customizable KPI, which are useful for the management staff in taking proactive decisions. The term was defined by Gartner Inc. (http://www.gartner.com/it-glossary/bam-business-activity-monitoring) and refers to the real-time aggregation, analysis, and representations of the enterprise data (possibly relating it to the system stakeholders and the customers).

The BAM's target is to produce (near) real-time information about the status and the outcome of the operations, processes, and transactions of a jBPM business system; this supports the corporate management staff in taking reactive decisions (**Decision Support System (DSS)**), and it helps staff to identify critical areas (possible sources of problems).

Examples include the following:

- Enterprises with a JIT production business model must constantly monitor their manufacturing and procurement processes and relate them to incoming orders and business providers

- Telco companies need to overview their services, providing operations in order to have an up-to-the-minute view of their customers

BAM typically needs to be integrated with BI/data warehouse tools; the first breed of tools is real time (data-oriented heterogeneous sources), while the second is historical business data. With the advent of NoSQL database engines, big data, and cloud-based platforms, this trend is today rapidly shifting away and turning to a new breed of tools handling streaming processing (real time) as well as batch processing (**Complex Event Processing (CEP)**).

BPM and BAM

The primary jBPM source for BAM data is the engine audit service and the jBPM database tables.

Audit data may be relevant for some business systems and useless for others. The auditing and logging history data could be a demanding task for your system/platform software and be very expensive in terms of I/O and/or allocated resources (disk space, DB resources, and so on). The jBPM audit logging service database schema is just a default implementation; the type and amount of the default audit data may not meet your needs, and your business application might require a finer (or just different) level of information to be captured.

The jBPM audit service module (`jbpm-audit-6.2.0.jar`) provides the implementers with two ways to produce audit data by collecting the engine events:

- **JPA**: Synchronous logger that is bound to the engine transaction and persists audit events as part of a runtime engine transaction

- **JMS**: Asynchronous logger that can be configured to place messages on the queue either with respect to active transaction (only after the transaction is committed) or directly as they are generated

[Please refer to *Chapter 6, Core Architecture*, for a thorough explanation of the jBPM auditing and logging services.]

Carefully evaluate the impact of the required audit service granularity on the engine performance and plan your implementation accordingly; for production environments, consider the following:

- Use a different database from the engine DB for your auditing data; this facilitates the DB management tasks (for example, no Foreign Key issues)
- Use asynchronous event processing for better throughput

Where audit data really matters, some systems typically require the process history log to be maintained for a considerable amount of time (years); planning a reliable database strategy (backup/recovery) is a must. The following list provides you with some first-hand tips:

- **No audit data required**: Turnoff JPA audit persistency options (configure the jBPM `persistence.xml` file)
- **Default audit data**: Enable JPA audit persistence
- **Custom/finer level but no extra processing required**: Turnoff JPA audit options, and follow the custom audit implementation notes in *Chapter 6, Core Architecture*
- **Custom/finer level and extra processing required (BI)**: Turnoff JPA audit options, and follow the custom audit implementation notes in *Chapter 6, Core Architecture*; writing to a different database is suggested

Please check *Chapter 6, Core Architecture*, for implementation details and samples.

Default history logs

The engine audit log/history information is stored in the PROCESSINSTANCELOG, NODEINSTANCELOG, and VARIABLEINSTANCELOG tables.

[Please check *Chapter 8, Integrating jBPM with Enterprise Architecture*, of the jBPM 6.1 user guide for details regarding the jBPM database schema.]

Lesser-known yet very handy tables for our BAM purposes are the TASKEVENT and AUDITTAASKIMPL tables.

The TASKEVENT and AUDITTAASKIMPL tables are managed by org.jbpm.services. task.audit.JPATaskLifeCycleEventListener, a task life cycle listener created and attached to the TaskService instance during runtime when JPA EntityManager is set.

The TASKEVENT table logs the task event transitions:

```
STARTED, ACTIVATED, COMPLETED, STOPPED, EXITED, FAILED, ADDED,
CLAIMED, SKIPPED, SUSPENDED, CREATED, FORWARDED, RELEASED, RESUMED,
DELEGATED, NOMINATED
```

Each row contains the key to the parent task (the **TASKID** column); you can take a look at an example full table dump file (taskevent_dump.txt) placed in the pizzadelivery Eclipse project. In the following paragraph, we will cover the BAM dashboard and build a customization example by using our PizzaDelivery business process audit data.

BAM and Dashbuilder – the pizza maker's performance index

jBPM 6 is shipped with a jBPM pre-configured web application built with Dashbuilder.

> Dashbuilder is the Red Hat open source platform for building business dashboards and reports; at the time of writing this book, the latest dashboard release is 6.2.0; please refer to the following product home site:
>
> http://www.dashbuilder.org.

The process and task dashboard is in no way a production-ready BAM platform but helps the jBPM system stakeholders (as well as the jBPM administrators) to get a consistent preview of the underlying auditing default data at nearly no cost.

> Please check Chapter 16 of the jBPM 6.2 user guide for an introduction to the jBPM dashboard.

To illustrate a hands-on sample case, we will now generate some audit data and add a new chart (backed by a new data provider) on our jBPM dashboard; the chart will display the performance index of all the process actors with respect to the assigned tasks.

Example projects

The book ships with several example Java projects; starting from this paragraph, we will use Maven projects that you can load and run into Eclipse without deploying them through the KIE workbench. The examples are jUnit-based test classes with single or multiple test methods each. To resolve all the required jBPM library dependencies, you can add the `org.jbpm:jbpm-test:6.2.0.Final` dependency to the `pom.xml` file generated by your Eclipse project.

Configuring persistence

The examples use Bitronix and the H2 database for persistence; the database connection settings can be configured for all the projects in the `localJBPM.properties` file. This file is found in the `test-common` project of the *Chapter 5, BPMN Constructs* examples onward and in the `Pizza` project for this chapter, in the `examples` folder. The main settings are as follows:

```
persistence.datasource.user=sa
persistence.datasource.password=
persistence.datasource.url= jdbc:h2:tcp://localhost/~/jbpm-62
persistence.datasource.name=jdbc/localjbpm-ds
```

Please note that `persistence.datasource.name` must be equal to the persistence unit `jta-data-source` element in the project `persistence.xml` file:

```
<jta-data-source>jdbc/localjbpm-ds</jta-data-source>
```

Generating audit data – ProcessBAM unit test

We leverage the default jBPM JPA audit listener and generate some audit data by using our `PizzaDelivery` process. Import the Maven `Pizza` project into Eclipse and run the `ProcessBAM` unit test; this class manages five threads simulating the operations of the five process actors. It introduces some delay in the task completion just to get realistic audit data (the time is expressed in milliseconds; the test takes a couple of minutes to run).

The test also makes Luigi slightly slower than Mario in making pizzas, so we can bring this difference to notice when evaluating the KPI.

```
salvatore is executing task 'deliver the pizza'
salvatore is waiting for 2000 before completing task 'deliver the
pizza'
```

```
mario is done with work for now.
salvatore is done with work for now.
...all tasks completed.
```

A number of 10 `pizzadelivery` processes shall be created and all related tasks completed:

- 10 incoming orders managed by Nino
- 10 order assignments performed by Maria
- 10-pizza cooking shared (randomly) by Mario or Luigi
- 10 delivery assignments performed by Maria
- 10 pizza deliveries performed by Salvatore

Let us see the steps required to configure and create the dashboard chart starting from the auditing data.

Writing the business query

The query (H2 SQL syntax and functions) calculates the duration (from the STARTED to the COMPLETED task event transition, in milliseconds) for each task of the 10 completed processes.

```
SELECT te.id,te.type as taskevent, te.logtime as startdate, te2.type
as taskevent, te2.logtime as enddate,
TIMESTAMPDIFF ('MILLISECOND',te.logtime,te2.logtime) as elapsed,
te.userid, t.name as taskname
FROM TASK as  t
INNER JOIN TASKEVENT as te on te.taskid=t.id
INNER JOIN TASKEVENT te2 on te2.taskid=te.taskid and te2.logtime in

(select tetmp.logtime from taskevent as tetmp where tetmp.logtime>te.
logtime and te.taskid=tetmp.taskid)

where te.userid is not null and te.userid  <>''
```

Starting from this data provider query, we are going to define two KPIs:

- Total time spent on tasks by the user
- Total number of tasks performed by the user

Adding the data provider

Log into the KIE workbench (admin/admin) and head to **Business Dashboard** (in the **Dashboards** menu):

1. Select the **Administration | Data Providers** link from the left navigation menu pane.

2. Create a new data provider: **name**=pizzerianapoli, **type**=SQL query, and as the **Query** field, paste the preceding SQL query; on pressing the **Attempt data load** button, the following message should appear:

```
SQL provider data load configuration is correct!
Elapsed time to execute query: 5189 ms
Number of entries: 50
```

The **Number of entries:50** message confirms that the query result is correct: 10 processes with five tasks each gives 50 tasks. Click **Save** to confirm the data provider setting. Clicking on the **Edit data providers** button will open the data provider list of the defined columns:

Properties	Type	Name	
id	Numeric ▼	id	English ▼
taskevent	Label	taskevent	English ▼
startdate	Date	startdate	English ▼
taskevent	Label	taskevent	English ▼
enddate	Date	enddate	English ▼
elapsed	Numeric ▼	elapsed	English ▼
userid	Label	userid	English ▼
taskname	Label	taskname	English ▼

Data Providers
Administration > Data providers
Editing data provider pizzerianapoli properties

Save   Cancel

We are now ready to define the KPIs, but before that, we need a new blank page on the dashboard on which we are going to place the charts.

Creating a new dashboard page and the KPI panels

Let us create a dashboard page that is going to host the charts.

1. Create a new page by clicking the **blank page** icon next to the **Page** list in the dashboard top toolbar (see the following screenshot):

2. Set the page settings (check the following screenshot), and click the **Create new page** button:

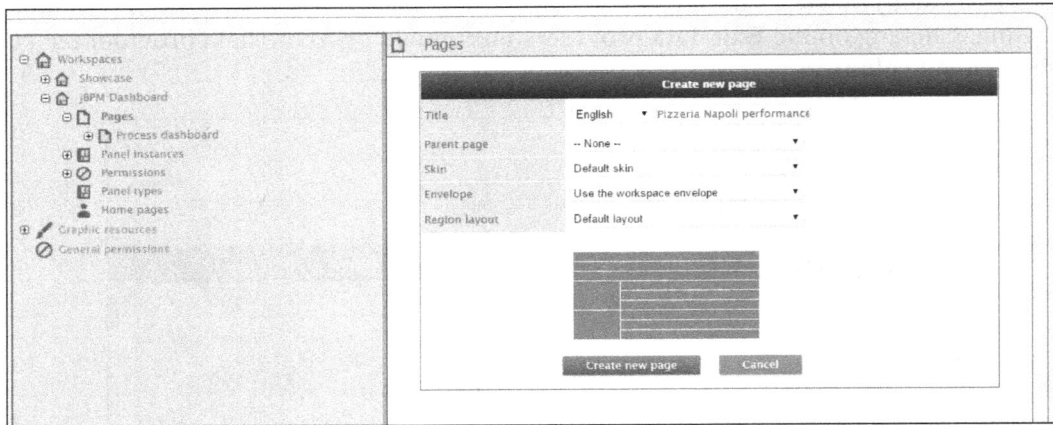

3. Click on the new page in the left navigation menu, insert the page URL in **Page properties**, and save changes.

4. Go back to the workspace and select the new page from the top page dropdown list; then, select the **Create a new panel in current page** item; a list with all of the available panels will pop up.

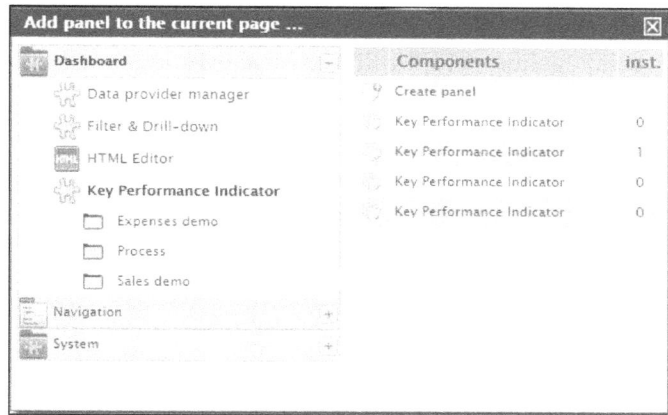

5. Drag the **Create Panel** item (on top of the **Components** list) and drop it on the highlighted target page panel that you wish your KPI to display; the dropped item prompts you for the source data provider. Selecting our `pizzerianapoli` data provider brings the KPI configuration panel to the front.

The relevant settings details for the KPIs are as follows:

Total time spent on tasks by the user KPI configuration:

- **Data provider**: pizzerianapoli
- **KPI name**: Total Time spent on Tasks by User (ms)
- **Bar Chart**
- **Domain (X Axis)**: userid
- **Range (Y Axis)**: elapsed; Edit Range (Scalar function: sum)
- **Renderer**: Open Flash
- **Chart type**: Box with perimeters

Number of tasks performed by the user KPI configuration:

- **Data provider**: pizzerianapoli
- **KPI name**: Tasks By User
- **Bar Chart**

- **Domain (X Axis)**: userid
- **Range (Y Axis)**: taskname; Edit Range (Scalar function: count)
- **Renderer**: NVD3

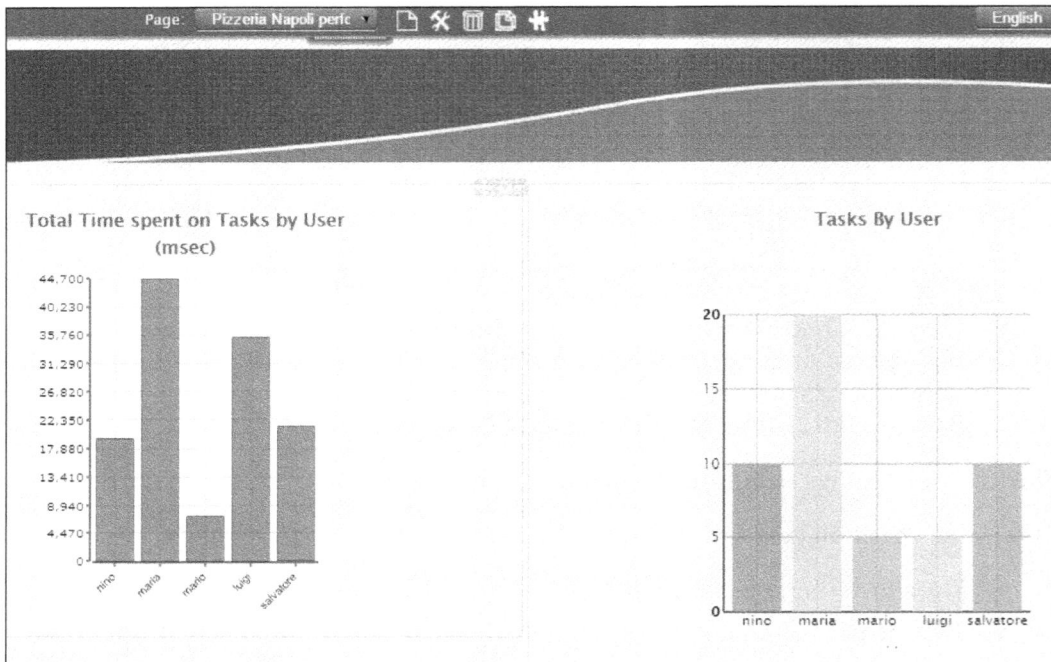

Notes

As we previously highlighted, the audit data elapsed times are denoted in milliseconds; this makes it possible to simulate random delays and avoids taking ages to run the test.

Given the randomness of the delays and the "make pizza" task assignments (both Mario and Luigi are potential owners of the same tasks), you might have different audit data; elapsed times and total tasks for Mario and Luigi will change for each test execution.

The unit test, however, always makes Luigi slower than Mario, so the overall KPI value never changes.

Of particular note in the two KPIs is the following:

- Mario is the fastest pizza maker
- Maria is the busiest employee (20 tasks: 10 order assignments + 10 delivery assignments); she must definitely improve her efficiency in order to not slow down the two pizza makers (waiting for her order assignments) and the pizzaboy (waiting for the delivery orders)

Summary

In this chapter, we explored the jBPM management features, discussing the new Git and Maven integration for project development and Kie module deployment, respectively. You should have a better understanding of the jBPM module management, and you should be able to create and configure BAM charts starting from the jBPM auditing data thanks to the off-the-shelf Red Hat Dashboard tool. The next chapter will dive into BPMN constructs and will provide you with practical process definition examples.

5
BPMN Constructs

To classify the level of support that a BPMN software tool provides, the BPMN standard defines the "conformance classes" as follows:

- **Process Modeling Conformance**: This class includes the BPMN core elements, process, collaboration, and conversation diagrams. It defines subclasses that contain a limited set of visual elements (descriptive), an extended set of modeling elements (analytical), and modeling elements that are required to model executable processes (common executable).

- **Process Execution Conformance**: It requires a software tool to support the operational semantics of BPMN.

- **Choreography Modeling Conformance**: The choreography modeling conformance class includes the BPMN core elements and the collaboration and choreography diagrams.

> jBPM supports a great part of the Common Executable class, with additional extensions. Please check *Chapter 6, Core Architecture*, of the jBPM 6.2 user guide for insights into the topic.

jBPM introduced the implementation of the BPMN 2.0 specification with the jBPM 5 release, for both the graphical notation (element visual representation) and the XML serialization, easing the task of exchanging process definitions between developers and the business team (in terms of Eclipse-based BPMN editor and process Web-based designer interoperability).

Other jBPM BPMN notable features are as follows:

- Compliance with the BPMN process execution semantics ("Common Executable" subclass specification)
- The BPMN **DI** (which stands for **Diagram Interchangeability**) specification for storing diagram information
- The BPMN I/O specification for input/output mapping

In *Chapter 1, Business Process Modeling – Bridging Business and Technology*, we already had an overview of the main BPMN concepts, constructs, and modeling patterns. We selected the topics for this chapter not to provide you with a BPMN modeling or reference guide, but as hands-on, example-driven explanation of all BPMN constructs supported by jBPM, without completely hiding away the underlying technical details.

In this chapter, we will discuss the following:

- The concept behind the BPMN construct
- How to use it in a business process (with examples)
- Best practices for when and where to use BPMN constructs

Parameters, variables, and data

Most of the time, business processes are data-driven processes: tasks handle variables, and rules handle facts; you will not be asked to draw a BPMN diagram without handling variables, parameters, objects, and states coming from external systems, user input, and other sources. A majority of the jBPM constructs are useless without data. Let us clarify the basics:

- **Parameters**: These are the data input coming from the user through the API. The user can pass parameters during process creation, at a human task completion, or into a service task for a Web service call.
- **Variables**: Variables are objects living in the scope of a single process instance. Variables can be created directly inside a process instance construct (for example, Script Activity and Data Object) or can be mapped from/to other variables (Data Input/Output Mapping) defined in another scope, for example, from the main process to a subprocess, from the process to a human task, and so on.
- **Globals**: Static variables shared across different process instances for a single Kie working session.

- **Facts**: Data that can be added to the Kie session and then updated or removed (retracted). This information is inserted, technically speaking, into the session through channels named **entry points,** and evaluated according to the Drools business rules, for activation. Drools Agenda manages the rule activation and firing mechanism.

> Please refer to Drools reference documentation for additional details on facts, rules, entry points, Agenda, and the Drools rule engine in general: `https://docs.jboss.org/drools/ release/6.2.0.Final/drools-docs/html`. Drools and jBPM are complementary projects that integrate together very nicely.

Variables and globals are accessed through context-type implicit references made available to the jBPM constructs at runtime:

- `ProcessContext (kcontext)`: This gives you access to variables
- `KieRuntime (kcontext.getKieRuntime())`: This gives you access to globals and facts

There are no implementation constraints on parameters, variables, and global class types apart from implementing the `java.io.Serialization` interface. Remember in fact that jBPM uses the standard in-memory serialization mechanism (`readObject/ writeObject`). When we enable persistence, it features an additional custom object marshalling mechanism to and from the store for session and process instances (see *Marshalling* in *Chapter 7, Customizing and Extending jBPM*). Furthermore, when there are persisting process variables for auditing and logging (`VARIABLEINSTANCELOG` table), jBPM stores the values by calling the process variable `toString()` method.

> jBPM does not provide out-of-the-box process variable persistence in any of its schema tables. We need to implement our ad-hoc variable serialization strategy (we will cover variables persistence with *Marshalling* in *Chapter 7, Customizing and Extending jBPM.*).

Sequence flow

The sequence flow is the connector between two elements of the process. It represents a flow of execution. A sequence flow may optionally have a condition defined (conditional sequence flow). The engine always evaluates a task node's outgoing sequence flows: If the condition evaluates to true then the engine selects and follows that sequence flow; a sequence flow with no condition defined is always followed by the engine. A **diamond shaped** connector (see *Appendix B, jBPM BPMN Constructs Reference, Gateways* section for some pictorial examples) indicates a conditional sequence flow. Multiple sequence flows represent branching and merging without the usage of a gateway. Gateways, depending on their nature, handle conditional sequence flows in specific ways as we are about to see.

> jBPM allows you to enable multiple outgoing conditional sequence flows from a task by setting the jbpm.enable.multi.con system property to true (default is false).

The following example process (see the figure) shows how the jbpm.enable.multi. con property affects the sequence flow behavior.

Example test class:

```
com.packt.masterjbpm6.SequenceTest
```

Example process:

```
sequenceflows.bpmn
```

Description: The test creates the process instance with an `Order` variable with different cost values. The process, thanks to the `jbpm.enable.multi.con` system property set to `TRUE`, allows the execution of multiple (here, we have two) conditional sequence flows that diverge from a single Script Activity. The first sequence flow is taken if the Order costs more than 10, while the second one is taken when the Order cost is ≤10.

Gateways

Gateways are elements that allow you to create branches in your process. These branches can be, conceptually, diverging or converging. You can model the behavior of the different types of business process sequence flows: conditional branching (inclusive and exclusive), forking, merging, and joining.

Let us first review the key gateway concepts and the practical examples in the upcoming sections:

- Fork (split) indicates a flow dividing into two or more paths that should execute in a logically parallel (concurrent) way: jBPM, for implementation reasons, never executes parallel flows concurrently (at the thread level) but always sequentially, one step at a time

- Join (or synchronization) refers to the combining of two or more parallel paths into one path

- Branch (or decision) is a point where the control flow can take one or more alternative paths

- Merge refers to a process point where two or more alternative sequence flow paths are combined into a single sequence flow path

Hence, the gateway **direction** property is defined as follows:

- **Unspecified**: May have both multiple incoming and outgoing connections

- **Mixed**: Multiple incoming and outgoing connections

- **Converging**: Multiple incoming connections and only one outgoing connection

- **Diverging**: Only one incoming connection and multiple outgoing sequence flows

Unspecified and mixed directions are not implemented

Let us now see how these BPM concepts translate into jBPM modeling elements.

Parallel (AND) gateway

This gateway allows us to fork into multiple paths of execution or to join multiple incoming paths of execution. When used to fork a sequence flow (diverging or AND-split), all outgoing branches are activated simultaneously. When joining parallel branches (converging or AND-join), it waits for all incoming branches to complete before moving to the outgoing sequence flow. This gateway must be used when many activities have to be carried out at the same time in any particular order.

Example test class:

```
com.packt.masterjbpm6.gateway.GatewayParallelTest
```

Example process:

```
gateway_parallel.bpmn
```

Description: The plan route script task calculates the order delivery route, while the **Prepare Ingredients** human task adds some mozzarella to the order bill of materials. The closing **Done** Script task displays the result after all outgoing flows are complete.

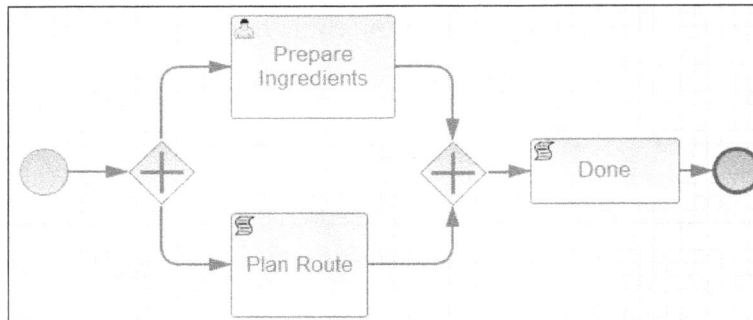

Conditional branching

These gateways introduce the *condition expression*. The condition expressions linked to each of the outgoing/incoming sequence flows are evaluated during process execution using process data (data-based gateways). Optionally, one of the gateway outgoing paths can be flagged as the **default flow** (its condition is ignored): this path is taken only if none of the other path flows can be selected. The default (sequence) flow is visually marked with a slash mark as shown in the following image:

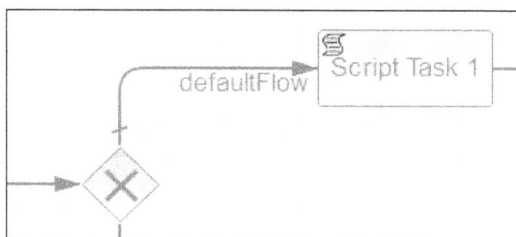

The "default flow" property is supported in the Exclusive and Inclusive Gateway elements.

Drools

We briefly introduced Drools facts and rules in the first section. Conditional branching based on Drools expressions works with facts but not with process variables. If we want to leverage the Drools expression features in the gateway constructs, we have to insert the process variable as a Drools fact, for example, given the process variable `order`:

```
Order order = new Order();
order.setNote("urgent");
order.setCost(110);
```

From inside the process definition (by a Script task, a Task **on exit** Script, and so on), we insert the following fact:

```
kcontext.getKnowledgeRuntime().insert(order);
```

Alternatively, we can do so by using the API as follows:

```
ksession.insert(order);
ksession.fireAllRules();
```

Exclusive (XOR) gateway

It is used to model a decision in the process. More than one path cannot be taken; the paths are mutually exclusive, hence, the name. In case multiple sequence flows have a condition that evaluates to true, the first one defined in the XML is selected for continuing the process. In an exclusive gateway, all outgoing sequence flows should have conditions defined on them. The default sequence flow is an exception to this rule.

Example test class:

```
com.packt.masterjbpm6.gateway.GatewayExclusiveTest
```

Example process:

```
gateway_exclusive.bpmn
```

Description: Different paths are taken for successful Pizza deliveries; the default path is chosen when other conditions are not met.

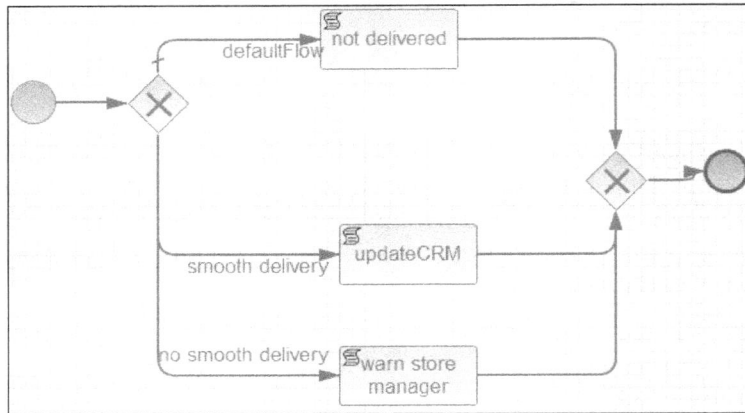

Inclusive (OR) gateway

An inclusive gateway is a branching point of the business process. Unlike the exclusive gateway, an inclusive gateway may trigger more than one outgoing flow and execute them in parallel (such as the parallel gateway). So, with diverging behavior, the gateway will always evaluate all outgoing sequence flow conditions, regardless of whether it already has a satisfied outgoing flow or not (unlike the exclusive gateway). In the case of converging behavior, the gateway will wait until all the incoming active sequence flows have reached it (merging). We can usually use this construct in a pair of splitting/merging gateways (see the following example) when we need to fork executions depending on certain conditions and then rejoin them.

Example test class:

```
com.packt.masterjbpm6.gateway.GatewayInclusiveTest
```

Example process:

```
gateway_inclusive.bpmn
```

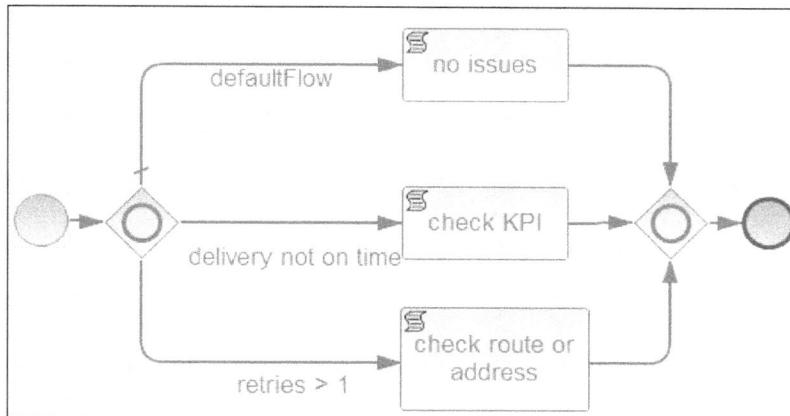

Description: Multiple different paths are taken for evaluation of the order delivery status; the testIssues test is set up so as to make the process take both the **delivery not on time** (deliveryDate > dueDate) and the **retries > 1** path. The default path is chosen when other conditions are not met (see the testNoIssues test).

Event-based gateways

Event-based gateways are similar to exclusive gateways, but the gateway trigger is based on event occurrence instead of condition evaluation. When our process arrives at an event-based gateway, we will have to wait until something happens. A specific event, usually the receipt of a message, determines the path that will be taken. Basically, the decision is made by another actor on the basis of data that is not visible to a process. This gateway is always a diverging gateway and must have at least one event attached.

Example test class:

```
com.packt.masterjbpm6.gateway.GatewayEventAndTaskTest
```

Example process:

```
gateway_event_and_task.bpmn
```

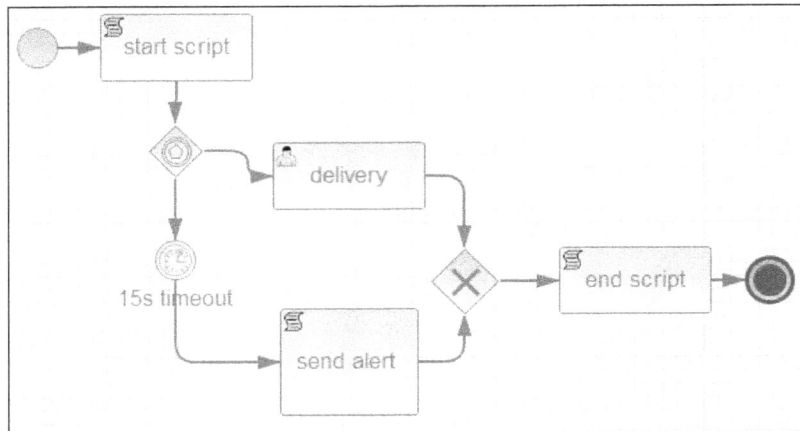

Description: The event gateway has a timer attached; when the timer expires, the **send alert** script is executed, bringing the process to termination.

Instantiating gateway

The instantiating gateway is a specialized event-based gateway, which triggers the process instantiation as soon as an attached event is received. The "instantiate" option (as of jBPM 6.2 the option is available in the jBPM Eclipse plug-in only) configures the gateway as a diverging gateway with no incoming connections: this gives you a way to instantiate a process by using an event, such as timer expiration or a catching signal event (see the following sections for timers and signals). jBPM does not support a pure instantiating gateway with no incoming connection: you always have to link it to a Start "None" event (see the following figure) or the process compilation will fail (complaining with a "missing incoming connection" error)

Example test class:

```
com.packt.masterjbpm6.gateway.GatewayEventTest
```

Example process:

```
gateway_event.bpmn
```

Description: Depending on events sent from an external (API call), different paths are taken (the `testCustomerPhoneCallEvent` and `testDeliveredEvent` methods); the timer triggers after 15 s if no event is caught (the `testTimerExpired` method). Note that both catching events pass the signal data (a randomly generated `orderid` string) to the process parameter `orderid`, which is later printed from the script tasks.

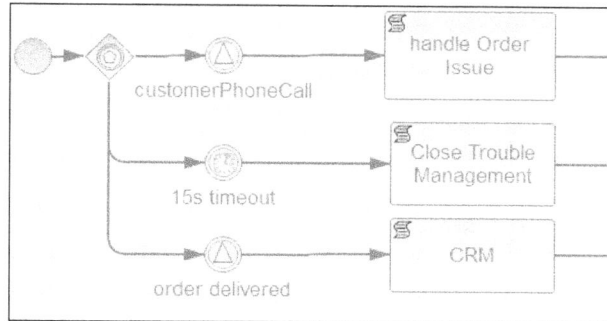

Complex gateway

This gateway can be used to model complex synchronization behavior. The construct options are available at the designer level, but jBPM has no implementation for this construct.

Events

Events are elements used to model something that happens during the process lifetime. BPMN 2.0 defines two main event categories: **catching** and **throwing** events.

- **Catching**: This event represents a pausing point in the process execution: Once the process flow reaches the catching event node, it stops in the wait state, waiting for a specific trigger to happen.

- **Throwing**: This event represents an action generating an event. When process execution reaches the event construct, an action is performed and a trigger is fired. For this throwing event, depending on the event type, there could be a matching catching event or not, that is, a send signal (throwing)/ catch signal or send error (throwing)/catch error. On the other hand, the compensate throw event does not have a catch companion, while the timer event is always a catching event.

Events are also categorized according to other criteria:

- An event can appear at the beginning of a process (Start event), within a process (Intermediate event), or at the end of a process (End event)

- An event can be generic or one of the different predefined types: time-based, message-based, signal-based, rule-based, exception-based, and so on

- An event can be positioned within a sequence flow or attached at the boundary of an activity (Boundary event)

- An event can exit the current process execution or not

> A note before we start:
>
> To facilitate reading, we'll go through the events by grouping them by event type (Start, Boundary, End) and then illustrating the supported variations (catching/throwing and start/intermediate/boundary/end) for each type of event (Signal, Message, Timer...).
>
> For additional information and a complete jBPM constructs reference (ordered the same way as you will find in both the Eclipse BPMN modeling tool palette and the KIE console palette), please refer to *Appendix B, jBPM BPMN Constructs Reference*.

Start events

The start event defines where (and how) the process is started; Start events are catching-only events. When a specific start event trigger fires (timer, messages, signal, and so on) the process is started. We will now see the None Start event; the other start event types are discussed in their respective sections.

Supported start events are: None, Message, Timer, Escalation, Conditional, Error, Compensation, Signal

None Start event

The simplest form of a Start event is the None Start event. It technically means that the trigger for starting the process instance is not specified; in other words, the engine does not know when the process instance is to be started. The only way to start the process is by invoking the `startProcess` method on a Kie session reference.

```
ProcessInstancestartProcess(String processId, Map<String, Object>
parameters);
```

End events

The End events are meant to express the end of a process or subprocess, and they are always throwing events. When the process execution arrives in the End event node, the associated event type is thrown. A process definition can have one or more End events defined. In this section, we will see the None and the Terminate End event; the other End event types are discussed in their respective sections.

Supported end events are: None, Message, Escalation, Error, Cancel, Compensation, Signal, Terminate

(None) End event

The None End event throws no events, and the engine just ends the current process instance sequence flow execution. If there are no more active sequence flows or nothing else to be performed (activities), the process instance is completed.

Terminate End event

The Terminate End event brings the process instance to the Completed state; all pending tasks, active sequence flows, and subprocesses are aborted.

Boundary events

Boundary events are events (always catching) that are graphically attached to an activity (subprocesses included) boundary (see the following figure). The event is registered for a certain type of trigger (see the following supported boundary events) and reacts only within the scope of the execution of the attached activity, with slight variations depending on the event type. In case the event triggers, it can optionally cancel the activity that it is attached to (by its cancelActivity property), and the event's outgoing sequence flow is executed. The boundary events are activated when the attached activity is started; in other words, they are bound to the activity instance life cycle. When the engine process execution path leaves the activity, all its attached boundary events are deactivated and their triggering is cancelled.

Supported boundary events are: Conditional, Error, Escalation, Message, Signal, Timer

See the *Boundary Message event* section for a working example.

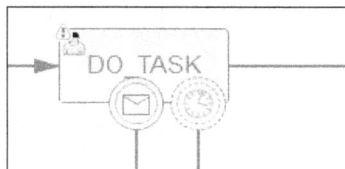

Signal events

A signal is a generic, simple form of communication, such as messages (see below). We can use signals to synchronize and exchange information. A catching signal may not have a corresponding throwing signal construct. It can also be sent programmatically from an external source (API). In contrast to other events (error event), if a signal is caught, it is not consumed. If there are two active intermediate catching events firing on the same signal event name, both events are triggered, even if they are part of different process instances and definitions. If the signal is sent and there are no catching signals registered for this event, the event is lost.

Scope

Signals can have visibility between different parts of the same process or broadcast processes (scope across all process instances), or targeted to a specific process instance. You can throw a signal event in a process instance, and other process instances with a different process definition can react to the event. Please keep in mind that this behavior (broader or narrower signal scope) can be affected by the *runtime strategy* chosen to create your Kie session (the subject is discussed in *Chapter 6, Core Architecture*).

Signal ID and signal name tips

You may notice some issues with signals when creating/modifying process signals in BPMN processes shared between the KIE jBPM console editor and the Eclipse BPMN modeler. The generated BPMN differs, and this may lead to bugs and unexpected behavior.

When creating the process definition from the Eclipse BPMN editor, the signal is assigned an internal ID of the form: `Signal_{number}`. Therefore, the actual signal ID to use is the same signal ID that you see in the `Signal` property editor and not the user-assigned signal name in the process definition panel (signal list table). Keep in mind this additional signal name referencing when coding against the `org.kie.api.runtime.KieSession.sendSignal` method.

```
<bpmn2:signal id="Signal_1" name="customerPhoneCall"/>
<bpmn2:signalEventDefinition id="SignalEventDefinition_1"
signalRef="Signal_1"/>
```

Therefore, with an Eclipse-generated process, the `Signal_1` ID must be used with the API.

```
<bpmn2:signal id="customerPhoneCall" name="customerPhoneCall"/>
<bpmn2:signalEventDefinition id="_05nSUW_YEeSWR_CUOywjGQ"
signalRef="customerPhoneCall"/>
```

With a process generated from a jBPM Web console editor, the signal ID is equal to the name attribute; customerPhoneCall must be used with the API.

Signal data mapping

Signals can carry optional object data; for each triggered catching signal, you can get this signal data and map it to a process variable. When operating with the jBPM Web designer, in order to successfully map the signal data to a process variable, you have to configure the **DataOutput** signal and assign it the name **event** as you can see in the following screenshot. The picture shows the event's data mapping for the start_signal.bpmn process signal events (see the *Start Signal event* section example for a detailed event data mapping example).

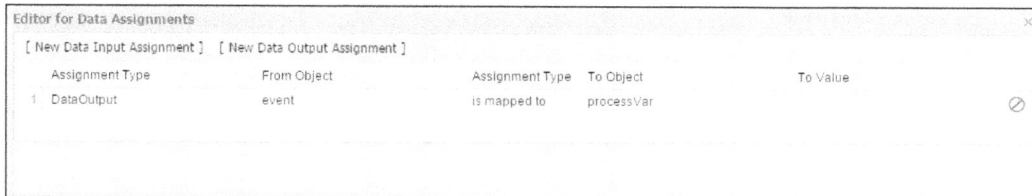

Editor for Data Assignments						
[New Data Input Assignment] [New Data Output Assignment]						
Assignment Type	From Object	Assignment Type	To Object		To Value	
1 DataOutput	event	is mapped to	processVar			⊘

This is a very flexible mechanism. By delivering data with your signals, you can update process variables, convey extra information, or change the process flow very easily.

Start Signal event

With a named Start Signal, we can programmatically start a process instance. The signal can be fired from within an existing process instance by using the intermediary signal throw event or through the API (the sendSignal method). In both cases, all process definitions that have a Signal Start event with the same name will be started. You can have multiple Start Signal events in a single process definition.

Example test class:

```
com.packt.masterjbpm6.event.StartTest (method testSignalStart)
```

Example process:

```
start_signal.bpmn
```

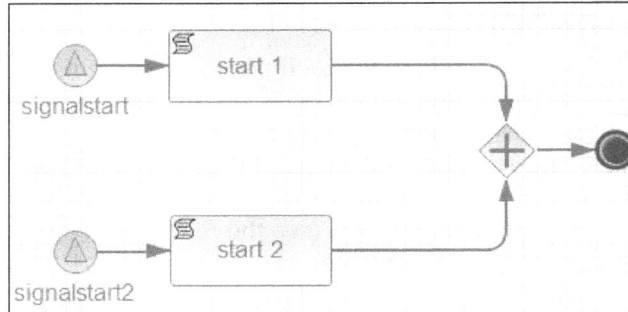

Description: Different Start Signal events are sent so as to create different process instances. The signal data is mapped to the process variable (see the previous section for an explanation of event data mapping).

Intermediate Signal event

An Intermediate catching Signal event catches signals sent from a throwing intermediate signal or through the API call (`KieSession` or `ProcessInstance.sendSignal`) and continues the process instance flow. The catching signal has no incoming connections.

Boundary Signal event

See the *Boundary events* section.

End Signal event

This kind of signal event is sent at the completion of the process. It can be a handy way to track process instance completions across the system.

Message events

Message events reference a name and can optionally have a payload. Unlike a signal, a message event is always targeted at a single catching message. The name of the catch and throw messages must be exactly the same in order to make the message flow work properly. Let us point out some differences between messages and signals:

- Inside the BPMN diagram, the message flow is drawn linking the sender to the receiver, while signals are never directly connected on the diagram. The throwing and the catching signal are implicitly connected only by their name.

- Messages should only be thrown/caught in the same process instance; there is no such limitation for signals. Messages work at the process instance scope only and are point-to-point links. A signal can travel from one process instance to many process instances (broadcast scope).

Message data mapping

See the *Signal data mapping* section.

Start Message event

A Start Message event is used to start a process instance as a direct consequence of catching a message; a process can have multiple Message Start events. This allows us to choose the process creation method simply by changing the Message event name to send (see the following image). Make sure that the message event name is unique across all loaded process definitions to avoid unwanted process creations.

> When sending the message from the API (`sendSignal`), we have to prefix the message name with the `Message-` string.

Message Start events are supported only with top-level processes and not with embedded subprocesses.

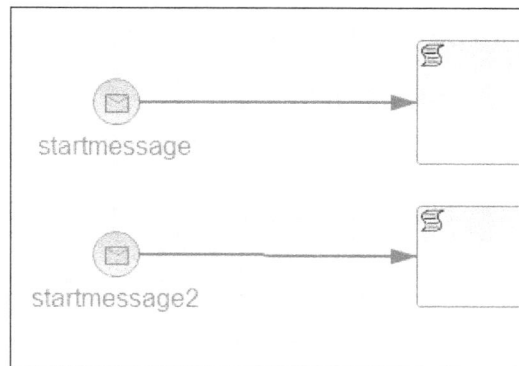

Intermediate Message event

If a process is waiting for the message, it will either be paused until the message arrives or change the flow for exception handling. For using a throw message, there has to be a catch message event that catches the message. It can be a message intermediate event or a message start event.

Boundary Message event

The following example shows task cancellation and message data passing by using two boundary events (a timer and a message) attached to a human task. The timer has the `cancel activity` property set to FALSE, while the message has it set to TRUE. The boundary message event maps the event data to a process variable in order to log the cancellation reason passed by the throwing (external) message sent by the test class.

Example test class:

```
com.packt.masterjbpm6.event.BoundaryTest (method
testBoundaryWithCancel)
```

Example process:

```
boundary.bpmn
```

Description: A process with a human task is created. The timer event's duty is to cycle and expire every 15 s calling the script task "time out warning" (its timer expression is `15s###15s`, and it is not flagged as "cancel activity"; therefore, the task will not be cancelled as the timer triggers). When the user continues with the test (the test class asks the user to press a key to proceed), a message is sent (`sendSignal`), the process message boundary event is triggered, and the activity is cancelled (since the boundary message event has the "cancel activity" flag enabled). Note that the message is sent by our test class with some data that serves as the task cancellation reason (`"cancelled by ADMIN"`):

```
sendSignal("Message-messageCancelService", "cancelled by ADMIN");
```

The boundary message (`id=messageCancelService`) catches the sent message, and the message event data, which is bound to the process variable **reason**, is printed in standard output by the **cancel log** script task.

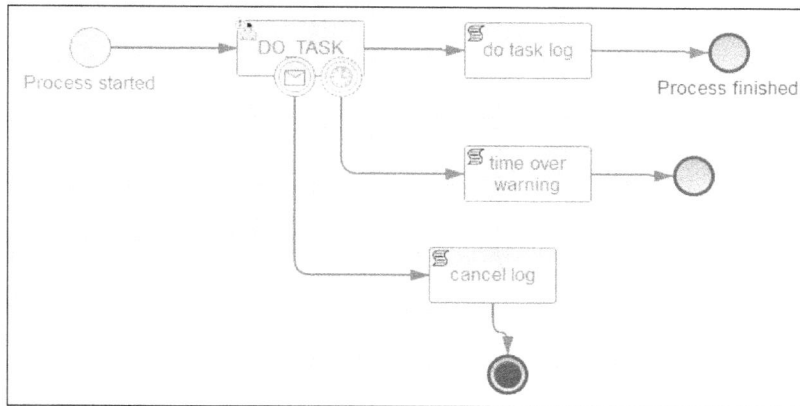

End Message event

A message is sent to a specific process at the conclusion of a process.

jBPM throwing message implementation

The jBPM throwing message default implementation is just a placeholder. You must provide your own WorkItemHandler definition and register it with the name Send Task to the jBPM Runtime, providing a hook to the working Kie session (identified by ksession in the following code fragment):

```
SendMessageTaskHandler messagehandler = new SendMessageTaskHandler();
messagehandler.setKnowledgeRuntime(ksession);
ksession.getWorkItemManager().registerWorkItemHandler("Send
Task",messagehandler);
```

> Throughout this chapter, you will find several references to "workItem" and "workItem handler and manager." These are the jBPM component part of a feature that lets you define a custom Java class and bind it with a specific process activity type in the engine runtime. Every time the engine activates this activity type, your handler will be invoked and passed the control. Please refer to *Chapter 7, Customizing and Extending jBPM* for detailed explanation and examples.

From the custom workItemHandler, you can then send signals:

```
public void executeWorkItem(WorkItemworkItem, WorkItemManager
manager) {
ksession.signalEvent("Message-startmessage", "processdata");
```

Example test class:

```
com.packt.masterjbpm6.event.StartTest (method
testMessageStartFromMessageThrow)
```

Example processes:

```
start_message_catch.bpmn, start_message_throw.bpmn
```

Description: The process created sends a message by a custom WorkItemHandler starting a new instance of the `start_message_catch` process (by a start message event).

Timer events

Timer events are events that are triggered when a timer construct expression is met; the timer properties are as follows:

- **Time Duration**: Single trigger delay value (for example: 10 m, 25 s).

- **Timer Cycle**: The time expression that shall be evaluated. It can be a string (interval-based 20 s or 5 m###35 s, where the first value is the initial delay and the second value is the delay between repeated fires), a string `cron` expression, or a process variable. In the case of JBPM 6.x, it can also be a ISO-8601 formatted date.

- **Timer Cycle Language**: Can be a default interval (empty value and time duration set) or `cron`.

Start Timer event

A Start Timer event is used to create a process instance at a given time. It can be used for processes that should start only once and for processes that should start at specific time intervals. Note the following points:

- A subprocess cannot have a start timer event.

- A Start Timer event is registered as soon as the process is deployed. There is no need to call the `startProcessInstance` API.

- When a new version of a process with a Start Timer event is deployed, the job corresponding to the old timer will be removed.

Intermediate Timer event

This event is a catching event only. The timer value triggers the execution of the outgoing sequence flow. You can use the timer to insert a generic delay or a timed-out sequence flow execution; for example, you could add a timer to manage a due date for a human task completion (see the *Event-based gateway* section example for a timer that acts this way).

Boundary Timer event

See the *Boundary Message event* section for an example.

Error events

Error events are used to model business exceptions. They are triggered by an exception that might be generated during the execution of an activity. Intermediary throw/catch error events do not apply.

Boundary Error event

This boundary error event must be attached to an activity. As the error event triggers, the activity is always canceled and the error event's outgoing sequence flow is taken.

Example test class:

```
com.packt.masterjbpm6.event.ErrorTest (method testBoundaryErrors)
```

Example process:

```
errorboundary.bpmn
```

Description: Two different boundary error events are attached to the same user task registered on different `errorCode` properties (`FileNotFoundException` or `RuntimeException`); the error handler logs the exception message. Depending on the process parameter (`triggerexceptionflag`) value passed, the user task throws a different exception upon completion (the `onExit` script), which triggers the appropriate boundary error event.

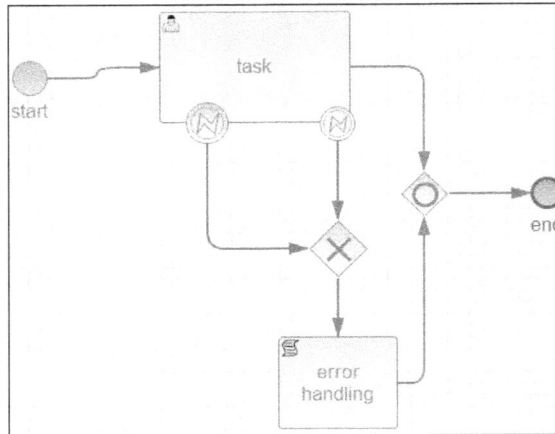

The process is started with a variable whose value affects the type of exception to be thrown:

```
Map<String, Object> params = new HashMap<String, Object>();
// "1" for a runtime exception; "2" for a FileNotFoundException
String trigger = "1";
params.put("triggerexceptionflag", trigger);
ProcessInstance processInstance =
ksession.startProcess("errorboundary", params);
```

The user task's `onExit` script evaluates the process variable and throws the exception accordingly:

```
String trigger=(String)context.getVariable ("triggerexceptionflag");
if (trigger.equals ("1"))
{
throw new RuntimeException("a runtime exception");
}
else
{
throw new FileNotFoundException("a filenotfoundexception exception");
}
```

The engine triggers the appropriate boundary error event depending on the exception thrown; the event, in fact, must be configured with the `errorCode` property set to the exception classname: `java.lang.RuntimeException` (see the following screenshot).

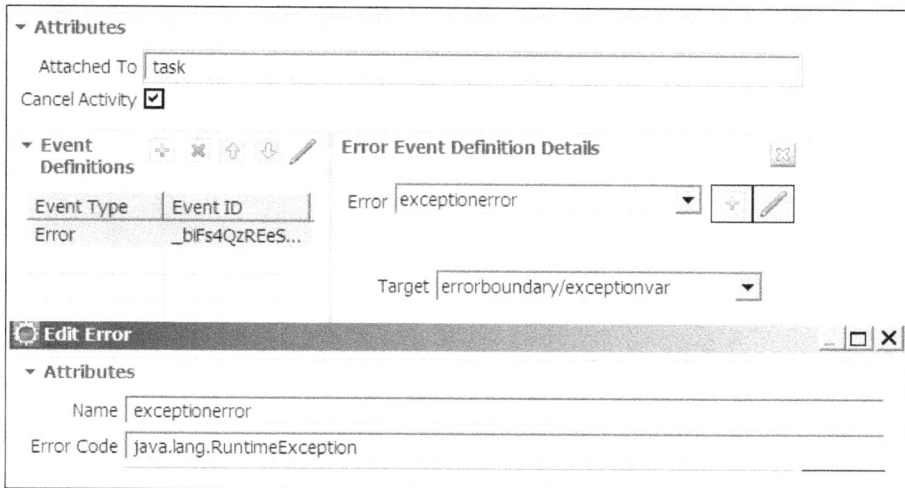

Note that the boundary error can bind the exception to a process variable. In the example, this variable (`exceptionvar`) is logged to the console by the script task:

```
Throwable exc=(Throwable )context.getVariable ("exceptionvar");
System.out.println("log error message:"+exc.getMessage());
```

Error Start event

The Error Start event can only be used to trigger an Event Subprocess and cannot be used to start a process instance. This is a feature you could consider using when activating alternative subprocesses on error exceptions.

Error End event

When the process execution reaches an Error End event, the current path of execution is ended and an error event is thrown. This error is caught by a matching intermediate boundary Error event or a subprocess Start Error event. If no Error event is found, an exception is thrown.

The following example uses the Error End event to trigger a subprocess by its Error Start event.

Example test class:

```
com.packt.masterjbpm6.event.ErrorEndTest (method
testSubprocessStartError)
```

Example process:

```
errorsubprocess.bpmn
```

Description: The main process features a human task and an Error End event, which triggers an embedded subprocess Script task by an Error Start event.

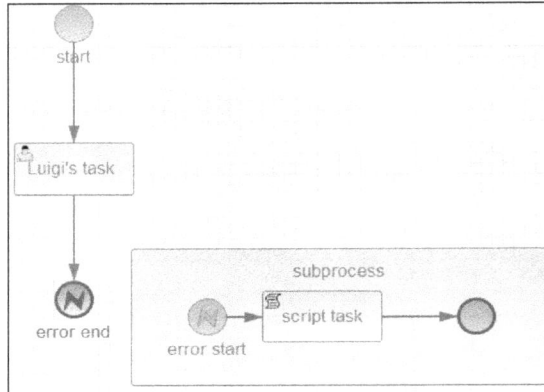

Compensation

Complex business processes may involve a number of heterogeneous parties and systems such as modern transactional systems, legacy systems (not transactional), and Web services. In order to preserve business consistency, when something fails and no transactional protocols are available, these systems may require you to perform programmatic corrective actions by invoking some dedicated API or by any other means. The compensation is the action of post-processing trying to remedy (not properly undoing or rolling-back) the effects produced by an action.

We want to stress the fact that jBPM compensations are not a transactional feature or a try/catch error mechanism. The compensation is a BPM business feature, which models an activity as the compensating counterpart for an already completed activity.

Here you have the common steps, which take place during a compensation event (see the following process example figure for a visual reference of the sequence).

- An activity (A1) whose boundary is attached to a compensation event (E1) is completed
- A compensate event (E2) is thrown somewhere in the process

- The compensate event (E1) catches E2
- jBPM activates the compensation handler (A2), which is connected to E1

The engine is ignorant of what the compensating activity will do since it is up to the developer to define the compensating business logic.

Intermediate Compensation event

The throwing Compensation event (E2) and the boundary Compensation event (E1) are implicitly connected by the same event name (we have already seen this with signals and messages). What we have explained for boundary events still applies here: when the Compensation event (E2) is triggered, the boundary Compensation event (E1) reacts by invoking the linked compensating activity (A2), which is marked with the typical compensation **FastBackward**-like symbol.

Boundary Compensation event

The Compensation boundary event (E1) must reference one Compensation handler (A2) only through the direct association line. The Compensation boundary event is activated only when the activity (A1) has been completed (unlike the default boundary event behavior where the event is activated depending on the Activity start state). The Compensation catch event (E1) is removed after either the parent process instance completes or the Compensation event itself is triggered. If a Compensation boundary event is attached to a multiple-instance subprocess, a compensation event listener will be created for each instance. jBPM does not seem to support this last feature.

Compensating activity

This activity (also called a compensation handler) is directly connected to the triggering boundary compensation event and must have no outgoing sequence flows.

Example test class:

```
com.packt.masterjbpm6.event.CompensationTest (method
testCompensationEvent)
```

Example process:

```
compensateorder.bpmn
```

Description: We used this example process to explain the typical compensation "workflow," so you should already be familiar with it. Let us just add that the Compensate event is thrown when the human task (**H1**) is completed and the `cancelOrder` variable evaluates to "y." This activates the exclusive gateway sequence flow, which triggers the event (**E2**). This activates the boundary Compensate event (**E1**), which in turn calls the **cancel order** script task (A2). The **cancel order** task acts as a "compensating" activity.

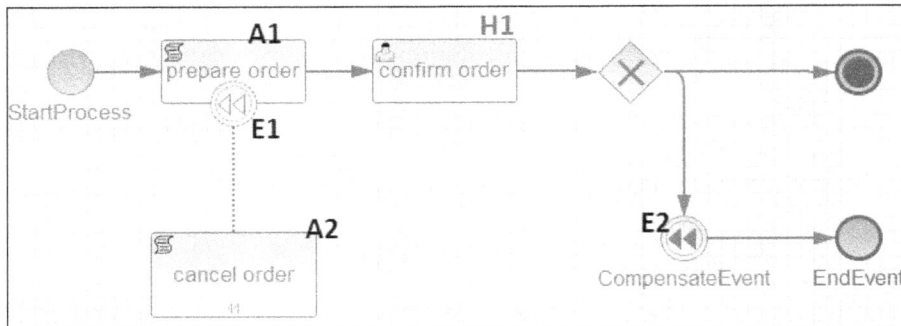

Triggering compensations with signals

jBPM offers additional ways to trigger compensations inside a process instance by using signals: general (implicit) and specific compensation handling. An **implicit** compensation triggers all of the compensation handlers for the process instance:

```
ksession.signalEvent("Compensation",
   CompensationScope.IMPLICIT_COMPENSATION_PREFIX
     + "compensateorder", pi.getId());
```

You must use the compensation signal type and pass the signal data a string that results from concatenating the `CompensationScope` class constant and the process definition ID resulting in the following:

```
"implicit:compensateorder"
```

The **specific** compensation triggers a specific compensation handler inside a process instance. You must pass the activity node ID attached to the boundary compensation event, along with the process instance ID:

```
ksession.signalEvent("Compensation", "_2", pi.getId());
```

Our example process script task XML element follows:

```
<bpmn2:scriptTask id="_2" name="prepare order"
scriptFormat="http://www.java.com/java">
```

> No new signal event needs to be defined at the process definition level.

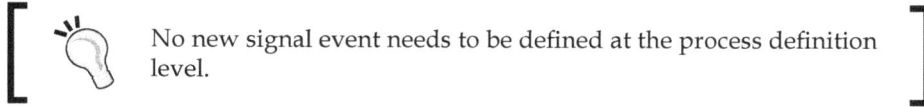

For working examples, please refer to the following:

Example test class:

```
com.packt.masterjbpm6.event.CompensationTest (methods
testGlobalCompensationWithSignal and
testSpecificCompensationWithSignal respectively).
```

End Compensation event

The end compensation event works the same way as the intermediate one (please see the example process figure). A compensation end event is thrown (**E1**), and the compensation handler triggered (**A1**). This kind of event is useful when there is a need to perform housekeeping or remediation business logic at the end of a process, but only when your bounded activity (**S1**) is in the COMPLETE state. Note in fact, as we already stressed, that the compensation handler kicks in only when the **subprocess** (**S1**) is already in the completed state.

Example test class:

```
com.packt.masterjbpm6.event.CompensationTest (method
testSubprocessCompensationEndEvent)
```

Example processes:

```
compensateendsubprocess.bpmn
```

Description: The process has a **subprocess** (**S1**) with an attached boundary Compensate event (**E2**). The subprocess triggers the throwing compensate end event (**E1**). The Compensate boundary catch event (**E2**) invokes the compensation handler (**A1**), which rolls back the process variable to the initial value.

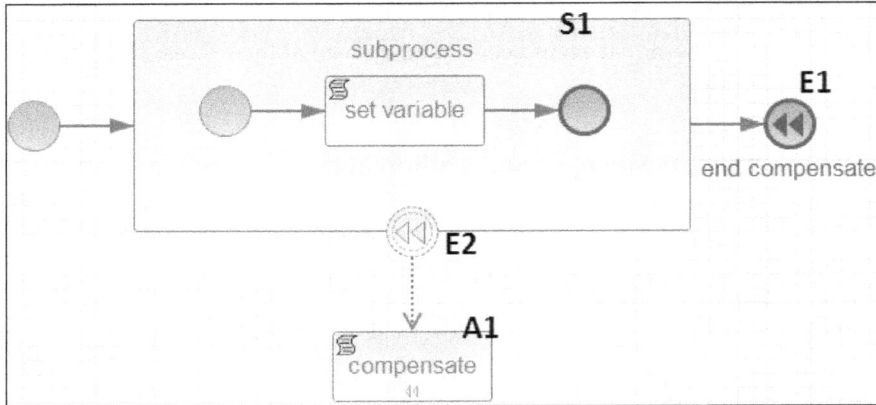

Multi-instance compensation

Compensation catching events attached to a multi-instance subprocess are not implemented. See the *Subprocess* section for details about multi-instance activities.

Escalation

Escalation, according to the common policies of an institution, organization, or corporate, refers to the existing relationships between the working personnel and their duties. The presence of an escalation event indicates that there is a condition that requires the business process flow to be diverted to a different user group. For instance, if an order above a certain price threshold is received, the approval task must be performed by a user in a higher role (for example, a manager); otherwise, it can also be approved by a clerk user.

In the case of jBPM 6.2.0, escalation events seem to be partially implemented and it is not clear what part of the BPMN specification is supported at this stage. You can partially overcome the lack of an escalation event with Deadlines and Notifications (see the *User Task* section).

Conditional events

Conditional events are a jBPM feature extension. They are triggered by an evaluation of user-provided expressions of Drools rules and facts properties. Conditional Start and Boundary events are supported.

Example test class:

```
com.packt.masterjbpm6.event.ConditionalTest (method
testSubprocessStartError)
```

Example process:

```
conditional.bpmn
```

Description: The main process is started when the Fact order note property matches "urgent"; the following script task **ordercost** is cancelled if Order cost > 100.

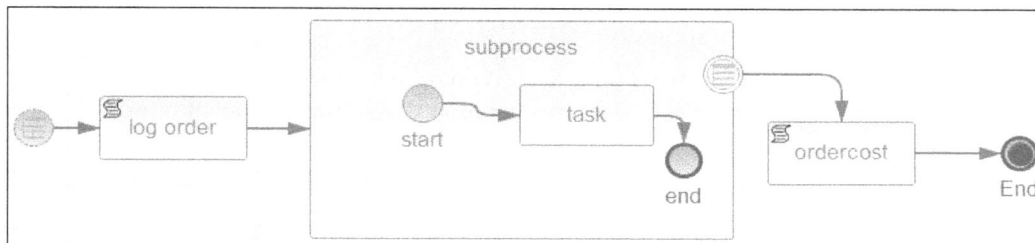

Activities

An activity is a unit of work that is executed within a business process; it can be atomic or non-atomic (Compound activity, Call activity, or Subprocess). Activities can be of the Task, Call activity, or Subprocess type.

Task

A task is the smallest atomic activity unit that can be included within a process. Usually, the performer of the task can be an end user (called human) using a UI-based application, a participating external service, or a generic set of business statements. Tasks have their local scope and can accept input parameters from their container and return output parameters.

User Task

A user task is used to model work that needs to be done by a human actor. When the process execution arrives at the user task node, a new task instance is created in the work list of the actor(s) or group(s) defined for this task (the Actors and Groups properties). Human tasks can transition to several different states and involve human stakeholders depending on the action taken on the task itself and the defined human roles.

Human roles

Human roles define what a person or a group of actors can do with tasks. Let us review the roles defined for the human task activities:

- **Task initiator**: The person who creates the task instance. Depending on how the task has been created, the task initiator may not be defined.

- **Actual owner**: The person who owns the task and is performing it. A task always has one actual owner.

- **Potential owners**: Persons who are given a task so that they can claim and complete it. A potential owner can become the actual owner of a task by claiming it.

- **Excluded owners**: Actors may not transition to be an actual or potential owner, and they may not reserve or start a task.

- **Business administrators**: Business administrators are able to perform the same operations as task users since they are always potential owners of every task. jBPM provides a default business administrator user (Administrator) and group (Administrators).

State transitions

The task remains in the **Created** state until it is activated. When the task has a single potential owner, it transitions into the **Reserved** state (it is assigned to a single actual actor); otherwise, it transitions into the **Ready** state; this state indicates that the task can be claimed by one of its potential owners. After being claimed, the task transitions into the **Reserved** state, elevating the potential owner to the actual owner actor. At this point, the actor can start the task that is in either the **Ready** or the **Reserved** state and make it transition to the **InProgress** state. The **InProgress** state means that the task is being worked on. If the actor completes the work, the task transitions into the **Completed** state. If the completion of the work goes wrong (exception), the task is put into the **Failed** state. Alternatively, the user can release the task, bringing it back to the **Ready** state. No transition is allowed from the **Complete** state and the **Failed** state.

> For detailed information on task state transitions, please refer to the Web Services – Human Task (WS-HumanTask) Specification by Oasis at `http://docs.oasis-open.org`.

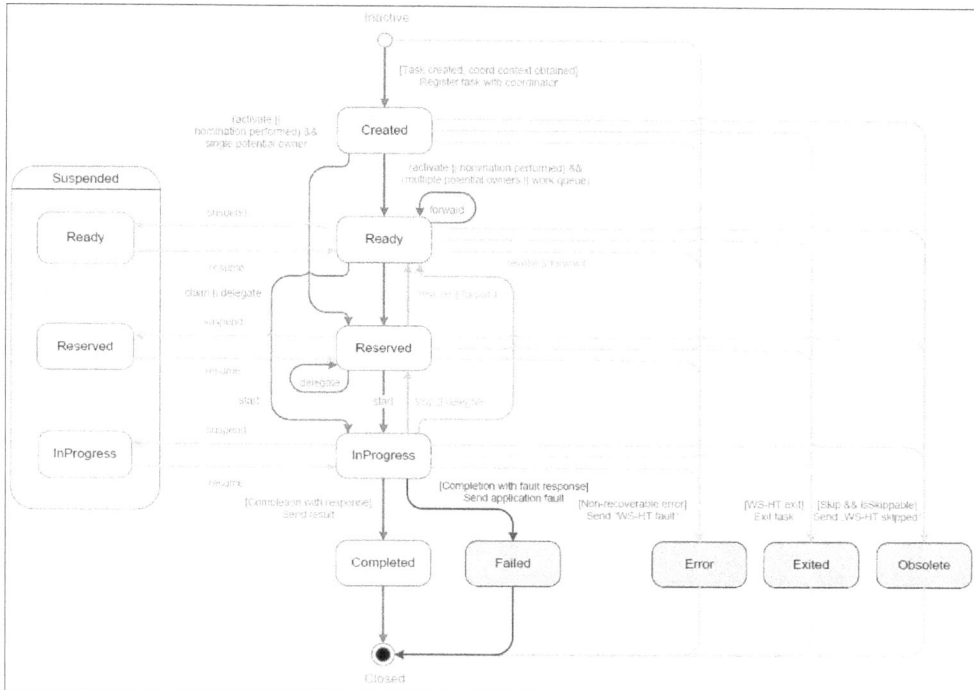

State transitions

Deadlines and escalations

The jBPM concept of a task deadline is bound to the task start-complete time interval duration; deadlines are associated with task escalations: the task escalation may exist in either a task reassignment or a task notification action. The task deadline is calculated on the task expiry date: it is reset when the task is started, and it expires when the task is completed over the allowed time boundary. Deadlines are physically stored in the DEADLINE table while notifications are stored in the NOTIFICATION set of tables.

> The Reassignment and Notifications property editor is available in the KIE Web process editor only.

Task reassignment

Task reassignment is a jBPM mechanism that lets you change a task ownership by setting specific rules, which are based on the task state transition and a deadline time expression, for example: "if Luigi (a named task actor) does not start the task in 60 seconds then reassign the task instance to Mario." The nominated user is replaced by the new user as the potential owner for the task. The resulting reassignment rule syntax is as follows:

```
[users:mario|groups:]@[60s]@not-started
```

You can define multiple reassignment rules on a single task instance.

Task event type conditions can be `not-started` and `not-completed`.

The BPMN XML task parameters are `NotStartedReassign` and `NotCompletedReassign`. The reassignment information is persisted by the engine into the `REASSIGNMENT` and `REASSIGNMENT_POTENTIALOWNERS` tables.

Example test class:

```
com.packt.masterjbpm6.activity.TaskTest (testReassign method)
```

Example process:

```
reassign.bpmn
```

Description: The main process is started, and the task is assigned to Luigi. The reassign rule states that "if Luigi (named task actor) does not start his task in 60 seconds then the task should be assigned to Mario."

Notifications

A notification is the action of alerting someone (actor, group) when a task deadline expires. The default jBPM notification is e-mail based, and the default e-mail configuration is read from the `userinfo.properties` and `email.properties` files. The `userinfo.properties` file lists the user/group information in the following form:

```
entityId=email:locale:displayname:[member,member]
```

e.g., for an entity of type actor, we have:

```
nino=nino@domain.com:en-UK:nino
```

Member data is optional and is used for listing members belonging to a group organizational entity.

> Please refer to the official jBPM 6.2 documentation for the configuration details.

The BPMN XML task parameters are `NotStartedNotify` and `NotCompletedNotify`.

An example `NotStartedNotify` parameter value follows:

```
from:mario|tousers:simo|togroups:|replyTo:|subject:warning|body:the
task has not been started in 10s !@10s@not-started
```

Delegation

Delegation is the process of setting a task's potential owners. The actual owners, potential owners, or business administrators can delegate a task to another user, adding this user to the potential owners (if he/she isn't already) and making the user the task owner. A task can be delegated when it is in an active state (Ready, Reserved, or InProgress) and transitioned into the Reserved state, and its `skippable` property can be flagged to `true` (the target actor/owner can skip the task). The task's state and parameters will not change after delegation.

Forward

Task forwarding is the process performed by a potential owner on an active task who replaces himself in the potential owner list, passing the task to another person. The potential owner can only forward tasks when in the Ready state. If the task is in the Reserved or InProgress state, the task is transitioned to the Ready state again.

Suspend/resume

A task can be suspended in any of its active states (Ready, Reserved, or InProgress), transitioning it into the Suspended state. The Suspended state has sub-states to indicate the original state of the task. When resumed, the task transitions back to the original state from which it had been suspended.

Skip

A stakeholder working on a human task or a business administrator may decide that a task is no longer needed and hence, skip this task. This makes the task transition into the Obsolete state. The task can only be skipped if this capability is specified during the task configuration (the `skippable` property).

For delegate, forward and skip, and suspend/resume examples have a look at a test class:

```
com.packt.masterjbpm6.task.TaskTest (methods
testDelegateReadyStateAndSkip, testForwardAndSkip,
testSuspendAndResume)
```

Example process:

```
delegate_forward.bpmn
```

Description: The main process is started, and a human task is reserved to Luigi. The test methods check for task delegation, forwarding, and suspend/resume.

Release

A task may be released by the current owner as a human task, making it available for other potential owners. From active states that have an actual owner (Reserved or InProgress), a task can be released and transitioned into the Ready state. Task data associated with the task is kept unchanged.

If a task is currently InProgress, it can be stopped by the actual owner, transitioning it into the Reserved state. Business data associated with the task as well as its actual owner is kept unchanged.

Script Task

A Script task is an automatic activity. When a process execution arrives at the Script task, the corresponding script is executed. All process variables that are accessible through the execution context (the kcontext variable) can be referenced within the script. It has the following properties:

- It is executed by the business process engine
- The script is defined in a language supported by the engine (Java or MVEL)
- The script task execution is always immediate
- The script task transitions to the complete state after the script execution

> For a complete MVEL reference, please visit http://mvel.codehaus.org/.

Example test class:

```
com.packt.masterjbpm6.activity.ScriptTaskTest
```

Example process:

```
script.bpmn
```

Description: The process script activity updates the process variable `order` description property:

```
Order order=(Order)kcontext.getVariable ("order");
order.setNote ("order modified");
```

Service Task

The service task indicates the work that is to be automatically performed by a service provider. Usually, all work that has to be executed outside the engine should be designed as a service task. jBPM supports two types of service task implementations: plain Java class and Web service. The service task is backed by a WorkItemHandler implementation (`org.jbpm.process.workitem.bpmn2.ServiceTaskHandler`) registered with the name **Service Task**.

The parameters are as follows:

- `Interface`: Java class name or WSDL WebService service interface
- `Operation`: Java method name or WSDL WebService operation
- `Parameter`: Method name (to invoke)
- `ParameterType`: Method (to invoke) parameter type (only 1 parameter supported)
- `Mode` (WS only): `SYNC` (default), `ASYNC`, or `ONEWAY`

In case of a service task of type Java, jBPM uses Java reflection to load the Java class type (by using an `Interface` parameter), instantiate it, and invoke the specified method (searched by `Operation` and `ParameterType`) with the value provided by `Parameter`. Only method signatures with a single parameter are supported, and the result of the invoked method is mapped in the activity `Results` output parameter.

The `Mode` parameter applies to a Web service only and describes the way a request has to be performed:

- **Synchronous (SYNC)**: Sends a request and waits for a response before continuing
- **Asynchronous (ASYNC)**: Sends a request and uses callback to get a response
- **Oneway**: Sends request without blocking (ignore response)

The Web service runtime leverages the "dynamic clients" features of the Apache CXF framework in order to generate Java classes at runtime.

[☒ Please visit `http://cxf.apache.org/docs/dynamic-clients.`
`html` for the official reference documentation.]

A Service task can be really useful for rapid prototyping, but when it comes to complex external service integration, it falls short in meeting common development needs: multiple parameter passing, additional Web service configuration, and so on.

The following example demonstrates how to override the standard jBPM service task component by adding a custom workItem handler. Note, however, that the input/output parameters of the custom service task handler cannot be changed from the process designer because the task interface is defined in the configuration files of the jBPM workItem handlers.

[☒ WorkItem handlers are thoroughly explained in *Chapter 7, Customizing and Extending jBPM*.]

Example test class:

```
com.packt.masterjbpm6.test.ServiceTaskTest
```

Example process:

```
servicetask.bpmn
```

Description: The first test (`testJavaServiceTask`) launches the process with a standard Java Service task (Interface: `ServiceJavaTask`, Operation: `processOrder`, Parameter: `order`, ParameterType: `Order`). The Service task changes the note field of the order and returns it to the main process whose script activity traces the change to the console. The second test (`testJavaCustomServiceTask`) features a custom Service task handler (`PacktServiceTaskHandler`) that overrides the default handler and processes the order parameter, setting its `note` property with a specific value.

Rule Task

The (Business) Rule tasks let us execute rules and get output from the embedded rule engine (Drools). Remember that process variables can be shared with the Rule tasks by using global variables or Drools session facts.

Example class:

```
com.packt.masterjbpm6.task.RuleTaskTest
```

Example knowledge artifacts:

```
rule.bpmn, rule.drl
```

Description: The main process is started, and the rule task triggers when the order cost is >100, and as a result, it changes the order's `note` property to URGENT. Look at the `rule.drl` file:

```
global StringBuffer newnote;
global com.packt.masterjbpm6.pizza.model.Order orderglobal;

rule "checkorder" ruleflow-group "masterRuleGroup"
    when
        $o: com.packt.masterjbpm6.pizza.model.Order (cost>100)
    then
    {
      System.out.println ("checkorder triggered");
      String desc="big order ! (cost="+$o.getCost()+")";
      orderglobal.setNote("URGENT");
      newnote.append (desc);
    }
End
```

The `order` variable (with `cost > 100`) is inserted into the knowledge session to activate the rule that triggers when `Order (cost > 100)`; see the `RuleTaskTest.testRule()` method:

```
ksession.insert(order);
```

While the shared `orderglobal` variable is used to get the result back:

```
ksession.setGlobal("orderglobal", order);
```

Send/Receive Task

Send/Receive tasks are general-purpose messaging tasks since they do not provide a default implementation. They are handled as workItem and it is up to the implementer to back them with a working implementation through the `WorkItemHandler` interface, registering it with the jBPM `WorkItemManager`.

The workItem name of the receive task must be **Receive Task**. **Receive Task** refers to the message ID through the `messageRef` attribute; the handler receives the message ID value with the `MessageId` parameter.

The workItem name of the send task must be **Send Task**. **Send Task** refers to the message ID through the `messageRef` attribute; for additional reference, check the Intermediate Message event.

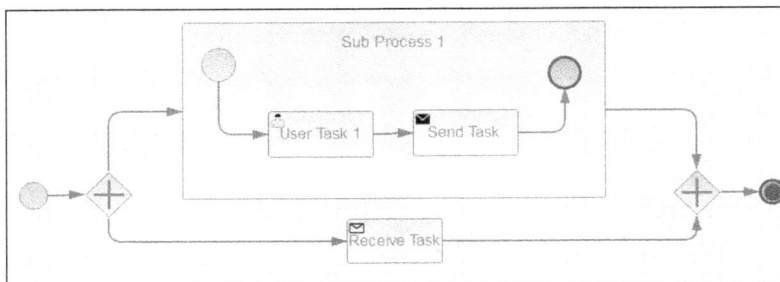

Example class

```
com.packt.masterjbpm6.task.TaskTest (method testSendReceive)
```

Example process artifacts:

```
send_receive.bpmn
```

Description: The subprocess send task passes data to the receive task of the parent process. The test registers two custom workItem handlers, and the Send task and the Receive task share a message by using a global process variable.

Manual Task

A manual task defines a task that is to be performed externally to the engine. It is used to model work that is done by a stakeholder without interacting with the system; the engine does not know anything about the task, and it does not need to. There is no UI interface or system available for the manual task completion. For the engine, a manual task is managed as a passthrough activity. It continues the process from the moment process execution arrives into it.

Ad hoc (Custom or None) Task

The custom task is an empty, generic, unspecialized unit of work. The implementer is requested to provide a WorkItemHandler implementation for the task and register it with WorkItemManager

```
Void registerWorkItemHandler(String workItemName, WorkItemHandler
handler);
```

> See *Chapter 7, Customizing and Extending jBPM* for detailed sections on the WorkItemHandler architecture.

The handler is registered for all workItems of the given workItemName and is called every time the process activates a node with that name. Further, workItemName must match the `taskname` attribute of the task element. WorkItemHandler is responsible for completing or aborting the task instance.

See the *Conditional* events section for a working example.

Async tasks

We are now going to take a closer look at some peculiar usage of the custom task. In *Chapter 4, Operation Management* we introduced the new jBPM executor service and the job scheduling features of the KIE console. The custom task can be conveniently configured to instruct the executor to call service-oriented components in an asynchronous fashion by scheduling an execution job in the background. The jBPM handler responsible for the job submission is `org.jbpm.executor.impl.wih.AsyncWorkItemHandler` (more on this in *Chapter 7, Customizing and Extending jBPM*).

> The jBPM process designer gives you the ability to toggle a `wait-for-completion` flag on the workitem handler node. This flag does not reflect the sync/async nature of the handler invocation. It does tell the engine to evaluate (by an event listener) the handler results and map them back to the process context variables, using the task output mapping. If the flag is set to false, the custom task results will be ignored.

We can configure an async task by doing the following:

- Specifying `async` as the task `taskName` property
- Adding a data input parameter called `CommandClass`, and assigning a fully qualified Java class name to the schedule
- (Optional) adding a data input parameter called `Retries`, which tells the executor how many times the execution should be retried (default = 3)

> *Chapter 4, Managing Jobs and Asynchronous Command Execution* explains in detail how to write `Command` classes.

The example that we discuss sets our `AsyncTaskCommand` as `CommandClass`, starts the executor service, and registers AyncWorkItemHandler.

Example class:

```
com.packt.masterjbpm6.task.AsyncTaskTest
```

Example process artifacts:

```
asynctaskprocess.bpmn
```

Call Activity Task

The call activity task is a general-purpose means to reuse existing, externally defined business constructs (process) simply by specifying their ID (the `calledElement` attribute of `bpmn2:callActivity`) or Name (`calledElementByName`). The execution of the called element can be synchronous/asynchronous (`waitForCompletion=true`) or independent (`independent=true`). You can set `independent` to false only if `waitForCompletion` is `true`.

All these properties are easily set, as usual, through both the jBPM Eclipse plugin or the KIE process editor; we extract from the process definition, for reference purposes, the relevant XML for the `callActivity` construct:

```
<bpmn2:callActivity drools:waitForCompletion="true"
drools:independent="true" name="CallActivity"
calledElement="callactivitySubprocess">
```

The following figure shows the main process on the left and the callactivitySub1 process "zoomed out" from the CallActivity node:

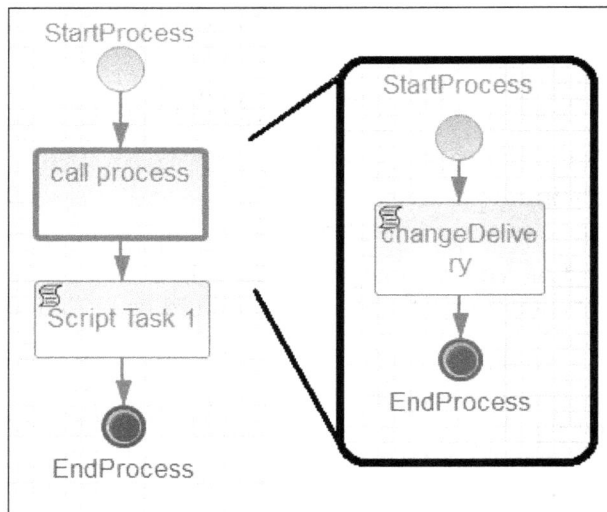

The callee construct supports, like other activity nodes (tasks), data input and output mappings from/to the caller process, as we are going to see in the following example.

Example class:

```
com.packt.masterjbpm6.task.CallactivityTaskTest
(testIndependentSubprocess method)
```

Example process artifacts:

```
callactivity.bpmn (parent process), callactivitySub1.bpmn (subprocess
called by the callActivity construct)
```

Description: The main process is started and callActivity is executed; the main process passes the process order variable to callActivity. The callActivity subprocess modifies the order variable and returns it to the calling process definition.

As a side note, if we examine the PROCESSINSTANCELOG table, we can see the two instances of the processes (the main and the called process) logged; their parentship relation is saved through the **PARENTPROCESSINSTACEID** column; it shows that **callactivitySubprocess** is a child process of the **callactivityprocess**. This is the output when callActivity has the independent=true and waitforcompletion=true properties set.

SELECT processid, processinsTANCEID , parentproCESSINSTANCEID , status , start_DATE , end_DATE FROM PROCESSINSTANCELOG:

PROCESSID	PROCESSINSTANCEID	PARENTPROCESSINSTANCEID	STATUS	START_DATE	END_DATE
callactivityprocess	1	null	2	2014-12-02 10:34:14.066	2014-12-02 10:34:14.253
callactivitySubprocess	2	1	2	2014-12-02 10:34:14.144	2014-12-02 10:34:14.191

Let us look at another example and see how the independent property affects the called subprocess.

Example class:

```
com.packt.masterjbpm6.task.CallactivityTaskTest (method
testAbortProcess)
```

Example process artifacts:

```
callactivityabort.bpmn (parent process),
callactivitysubprocessabort.bpmn (subprocess called by the call
activity construct)
```

Description: The `callactivityabort` process is started, and callActivity (with `independent=false`) is executed. The subprocess referenced by callActivity (`callactivitysubprocessabort`) has a human task, so it stops for user interaction. This gives us the time to issue (see the test class code) `abortProcessInstance` on the parent process. The `independent` flag set to `FALSE` forces callActivity (that is, the waiting subprocess) to abort contextually to the main process instance; when the flag is set to `TRUE`, the callActivity is not affected (see previous example).

This is the output when aborting the parent process instance, which has callActivity with the `independent=false` property set. Note also that `status` = 3 (ABORTED) for both process instances.

SELECT PROCESSID,PROCESSINSTANCEID , PARENTPROCESSINSTANCEID , STATUS, START_DATE , END_DATE FROM PROCESSINSTANCELOG;					
PROCESSID	PROCESSINSTANCEID	PARENTPROCESSINSTANCEID	STATUS	START_DATE	END_DATE
callactivityabort	1	null	3	2014-12-02 13:12:47.199	2014-12-02 13:13:00.57
callactivitysubprocessabort	2	1	3	2014-12-02 13:12:47.277	2014-12-02 13:13:00.507

Subprocess

A subprocess, as the name suggests, is a process that is included within another process. It can contain activities, events, gateways, and so on, which form a **boxed** process that is part of the enclosing process. The subprocess can be completely defined inside a parent process (an embedded subprocess) or can be linked through a CallActivity element by its ID or Name property. You can link a subprocess (by callActivity) across different multiple process definitions, reusing common groups of process elements (activities, gateways, and so on). The embedded subprocess construct can have multi-instance capabilities (see the MultiInstance section). However, using a subprocess does impose the following constraints:

- Sequence flow cannot cross subprocess boundaries
- Process variables must be mapped for input and/or output

At the designer level, a subprocess can be expanded or collapsed so as to hide or show its details.

Ad hoc subprocess

Ad hoc subprocesses are commonly used when a number of tasks can be selected and performed in any order (because unspecified or unknown), and there is no execution dependency between them. Tasks might have unknown dependencies, most often because they are dynamic and managed by a human user on a case-by-case basis. The subprocess can complete even if some of the tasks are not executed at all. An ad hoc subprocess is represented as a subprocess with a tilde (˜) marker at the base.

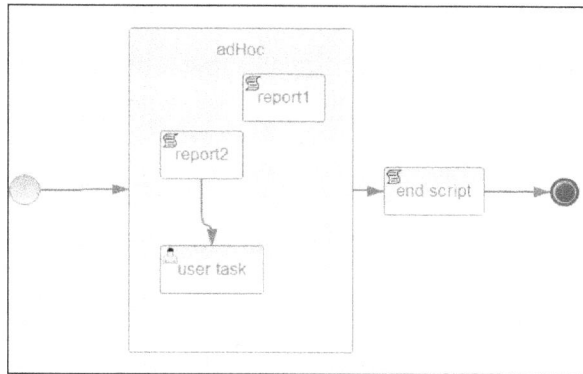

The jBPM ad hoc subprocess implementation seems to be fairly incomplete. There seem to be some issues when exiting from the subprocess instance. The user is able to start the ad hoc subprocess activities by using the `signal` method by referencing the activity name:

```
ksession.signalEvent("report1", null, processInstance.getId());
```

Because of their nature, ad hoc subprocesses are hard to design and of little use in real structured business processes; nevertheless, here, we provide you with an example that you can tweak and experiment with:

Example class:

```
com.packt.masterjbpm6.task.AdHocSubprocessTest
```

Example process artifacts:

```
adhocsubprocess.bpmn
```

Description: The ad hoc subprocess has 2 script activities and 1 human task. The script tasks are signaled, and the human task is completed.

Multiple instances

This construct can be used to create multiple instances of a reusable subprocess definition as well as an embedded subprocess. Passing an input parameter collection works as the instantiation loop. jBPM will create one instance of the looping process for each element in the collection. The following figure shows the process with the embedded multi-instance subprocess (**Log pizzas**, the parallel symbol denotes that it is a multi-instance process) and the subprocess attributes. The loop input is the process variable `list` and the loop instance parameter (the collection item) is `item` of type `Pizza`. The `item` variable is visible in the instantiated subprocess scope.

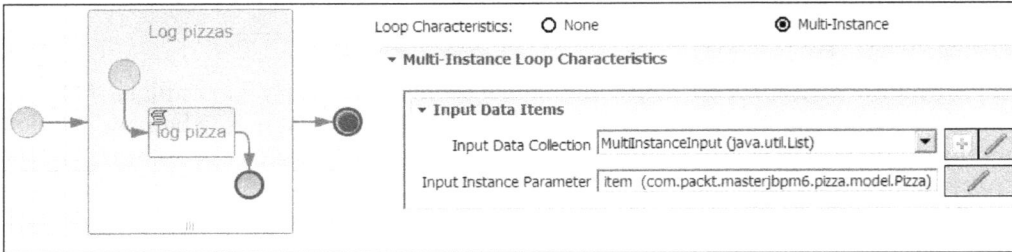

Example class:

```
com.packt.masterjbpm6.task.MultiInstanceTest
```

Example process artifacts:

```
multiinstance.bpmn
```

Description: The process is created by passing a variable `list` of pizzas:

```
List<Pizza> myList = new ArrayList<Pizza>();
myList.add(new Pizza(PizzaType.getType(Types.MARGHERITA),
"margherita"));
myList.add(new Pizza(PizzaType.getType(Types.NAPOLI), "assorreta!"));
params.put("list", myList);
ProcessInstance processInstance =
ksession.startProcess("multiinstance", params);
```

Subsequently, two subprocess instances are created, and each is passed the loop `item` variable (a `Pizza` instance). The subprocess script activity simply prints the pizza description, and the subprocess exits.

```
System.out.println("pizza desc " + item.getDesc());
```

Lanes

A lane is a partitioning box-shaped element used to group activities within the process definition. Lanes can be used to visually point out different group task assignments. For example, you can think of a lane as a company department (IT, business administration, and so on) where all employees have (more or less) the same duties. jBPM will try to assign (making a task reserved for the user) all tasks within the same lane to the same user. For example, if there are several tasks on a lane, the user who claimed and completed the first task will be assigned to the other tasks on the lane. Usually, it is convenient to assign the same group ID to all the tasks in the same lane.

Example class:

```
com.packt.masterjbpm6.task.LaneTest
```

Example process artifacts:

```
lane.bpmn
```

Description: The **task1** and **task2** (on **lane**) activities are assigned to the `pizzerianapoli` group, while **Mario's Task** is assigned to the actor Mario. **taskNotInLane** is also assigned to `pizzerianapoli` but it's not on **lane**.

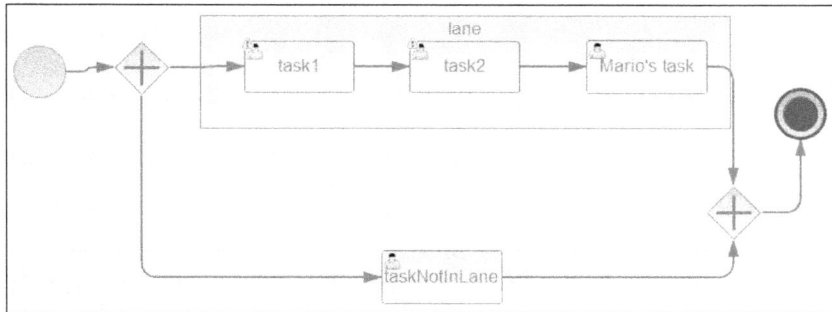

After the process is started, the actor Luigi (belonging to the `pizzerianapoli` group; see the `LaneUserCallBack` class) has 2 tasks on the list (**task1** and **taskNotInLane**). After he completes task1, he is automatically given the task2 activity (status = **Reserved**), while the **taskNotInLane** status remains unchanged (**Ready**).

SELECT name, status, actualOWNER_ID FROM TASK:

NAME	STATUS	ACTUALOWNER_ID
taskNotInLane	Ready	*null*
task1	Completed	luigi
task2	Reserved	luigi

Data objects

Data objects are BPMN constructs that represent how data is required or produced by an activity. Data objects can have a direct association to one or more activity providing the input or the target output for that activity.

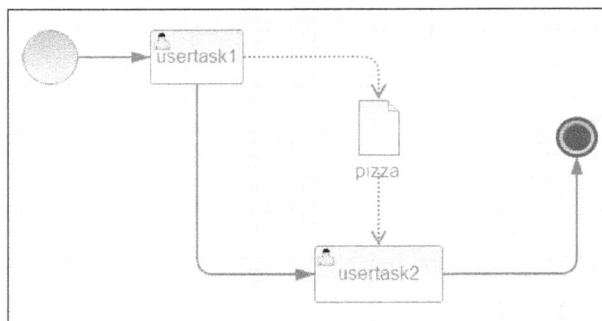

Example class:

```
com.packt.masterjbpm6.task.DataObjectTest
```

Example process artifacts:

```
data-object.bpmn
```

Description: The task1 and task2 activities share the same data object (pizza class type); the first task produces the pizza, which then serves as the input of the second task.

Summary

In this chapter, we examined the jBPM BPMN constructs, providing hands-on working examples, tips, and, whenever possible, some details regarding the jBPM internal mechanisms. The chapter is not meant to be a BPMN tutorial or a BPMN best practices modeling guide for which we suggest picking more suitable books and a lot of real-world practice. In the next chapter, we will cover the jBPM subsystems API with several practical examples: the new Kie API, the runtime engine, the human task service, and the persistence engine.

6
Core Architecture

Chapter 1, Business Process Modeling – Bridging Business and Technology, provided you with an overview of the new KIE platform and the jBPM tool stack components. This chapter will show you how jBPM is built and what its components and subsystems are, and it will take you through the source code of jBPM, illustrating, with examples, how to leverage the flexibility provided by its modular system.

The list of topics that we will cover in this chapter is as follows:

* Core API
* Runtime engine
* Human Task service
* Persistence and transaction
* History logs

The KIE API

The new KIE API stems from the need for providing both a new common software service layer and an integrated IDE (Workbench) for well-assessed Red Hat projects, mainly Drools and jBPM. With the KIE API, several features have been added to ease the integration of these platforms with several environments: JMS, Rest, SOAP, CDI, Spring, Seam, OSGi, and plain Java applications.

The **droolsjbpm-integration** additional project (hosted at `https://github.com/droolsjbpm/droolsjbpm-integration`) features integration packages for various environments and technologies.

We previewed some of the new KIE concepts in *Chapter 4, Operation Management* (new Maven-based deployments, KieModule, the `kmodule.xml` file, and KieScanner), so you should be ready to go deeper into the subject. You will also find, as a companion support for our examples, class diagrams of specific KIE component relationships, which should help you to have a clearer picture of the internal KIE organization.

[✎ `http://www.kiegroup.org` is the portal for all KIE technologies.]

KieServices

KieServices is a thread-safe singleton, which acts as a service factory. It gives high-level access to main KIE components and services.

It is possible to obtain a KieServices reference via its factory as follows:

```
KieServices ks = KieServices.Factory.get();
```

The main KIE API services are as follows:

- **KieContainer**: This is essentially a wrapper for KieModule and KieBase(s) that it defines; it can compile and verify KieBase and generate new KieSession(s).

- **KieRepository**: This is a repository that manages KieModules abstracting away from the module source; it can be a module installed in a Maven repository or a module programmatically created and added by the user.

- **KieScanner**: This is a Maven repository artifacts scanner.

- **KieBuilder**: This is a helper for compiling and building a KieModule starting from its set of source files.

- **KieResources**: This is a factory for creating specialized resources from various I/O sources (Classpath, URL, and so on).

- **KieLoggers**: This is a logger configurator for the session.

- **KieStoreServices**: This is a store service that manages the persistency of the jBPM and Drools runtime state.

- **KieMarshallers**: Marshalling provides a customizable serialization architecture, utility classes, and strategies. We will describe the jBPM flexible marshalling architecture in *Chapter 7, Customizing and Extending jBPM*.

Let us start by discussing the KIE API components whose duties are related to the jBPM runtime configuration and setup, since all your knowledge artifacts represent, at runtime, the engine's building ground for your business process execution.

The upcoming sections are not logically grouped under this KieServices section since they all are KIE first class citizens (classes) and can be used and created irrespective of the KieServices factory class. You can find the complete source code examples for this chapter in the `jbpm-misc` project.

KieContainer – KIE modules and KieBase(s)

`KieContainer` has been specifically designed to handle a KIE module and resolve its dependencies (other KIE modules or Mavenized JARs), even through remote Maven repositories. This is a huge improvement in terms of knowledge module sharing and management capabilities, compared to older jBPM versions. While a KIE module is an assembly that collects a set (archive) of business artifacts and static resources, it is the responsibility of KieContainer to organize the KieBase and KieSession definitions and to give the user the tools to obtain new ready-to-use references for them.

You can find the examples in the `KieContainerTest` class.

`KieContainer` can use the Java Classpath or the user-provided ClassLoader to detect, load, and wrap an existing `KieModule`:

```
KieContainer kContainer = ks.getKieClasspathContainer();
```

The `getKieClasspathContainer()` function returns `KieContainer` that wraps `KieBase` found in your current Classpath (created by parsing the available `kmodule.xml` files).

In *Chapter 4 Operation Management*, we talked about the new Maven repository integration feature. `KieContainer` is able to load KieModule from a Maven repository, given its Maven **GroupId-ArtifactId-Version (GAV)**; use the `ReleaseId` class as follows:

```
// create the Maven GAV wrapper
ReleaseId releaseId = ks.newReleaseId("com.packt.masterjbpm6",
"pizzadelivery", "1.0-SNAPSHOT");
// then create the container to load the existing module
KieContainer kieContainer = ks.newKieContainer(releaseId);
```

The container, by putting the KIE client repository service at work, is also able to dynamically update its definitions starting from a different KIE module, given its Maven GAV. As a result, all its existing KieBase assets and KieSession definition will be incrementally updated (and the cached class definitions will be replaced with newer ones).

```
ReleaseId newReleaseId = ks.newReleaseId("com.packt.masterjbpm6",
"pizzadelivery", "1.1-SNAPSHOT");
// update the container with the KIE module identified by its GAV
Results result = kieContainer.updateToVersion (newReleaseId);
if (result.hasMessages (Level.ERROR))
{
List<Message> errors= result.getMessages(Level.ERROR);
```

KieBase (KnowledgeBase) is the building block of KieModule. The `KieBase` class works as the store for the KieModule knowledge definitions and serves as a dictionary for your KIE session. It contains Drools rules, processes, models, and so on. By default, these artifacts are searched in the KIE project `resources` root folder, but you can set the `packages` attribute to search in a different folder, for example (an excerpt of a `kmodule.xml`):

```
<kbase name="kbase" packages="com.packt.masterjbpm6.event">
```

This will load artifacts from the `resources/com/packt/masterjbpm6/event` project folder.

Your KieModule must always have at least one named KieBase (that is, its `name` attribute must be set); alternatively, if you decide to use a `default` (that is, created without knowing its name) KieBase, omit the `<kbase>` element definition altogether in your `kmodule.xml` or leave the `kmodule.xml` empty.

> KieSession does not make much sense without an underlying KieBase. The KIE runtime, in case the user is not specifying one, provides you with a default KieBase. This default KieBase is KieBase with the attribute `packages="*"`, meaning that it is defined with all assets contained in all module packages.

KieBase is created by KieContainer and supports inheritance (inclusion) and multiple definitions:

- **Inclusion**: All knowledge artifacts belonging to "included KieBase" are added to "including KieBase"; for example, all `kbaseold` resources are added to the `kbasenew` KieBase:

```
<kbase name="kbase" includes="kbaseold"
packages="com.packt.masterjbpm6.process">
    <ksession name="ksession" />
</kbase>
```

The included KieBase must be already available (the KieModule within which it is defined has to be deployed) or defined locally (the same `kmodule.xml` file). In *Chapter 4*, *Operation Management* (the ManagedVesuvio repository example), the Napoli KieModule's kbase is included the Vesuvio's kbase in order to reuse its external process definition as a subprocess; let us clarify by looking at their Kie module definitions.

The Napoli `kbase` definition (relevant part only) is as follows:

```
<kbase name="kbase-napoli" default="true" packages="*"
includes="kbase-vesuvio">
```

The Vesuvio `kbase` definition is (relevant part only) as follows:

```
<kbase name="kbase-vesuvio" default="false" packages="*">
```

Note that, in order to have KIE pick up the main kbase (`kbase-napoli`), we set `kbase-vesuvio` kbase's `default` attribute to `false`.

- **Multiple definitions**: Multiple KieBase (and KieSession) can be defined inside a single KieModule:

```
<kbase name="kbase" includes="kbaseold"
packages="com.packt.masterjbpm6.process">
    <ksession name="ksession" />
</kbase>
<kbase name="kbaseold" packages="com.packt.masterjbpm6.event">
    <ksession name="ksession2" />
</kbase>
```

Once the KieBase is defined, you can create a stateful KieSession (the default one or a named one, by passing its `name` attribute).

```
KieSession kieSession=kieContainer.newKieSession ("ksession");
```

Each KieSession is always paired with a single KieBase: KieContainer actually delegates the session creation to its KieBase.

> KieBase and KieSession support a number of declarative configuration settings that you can add to your `kmodule.xml` file; please consult the jBPM 6.2 reference documentation.

The following class diagram shows the main classes that you have to deal with when working with containers (sessions will be discussed in a forthcoming section).

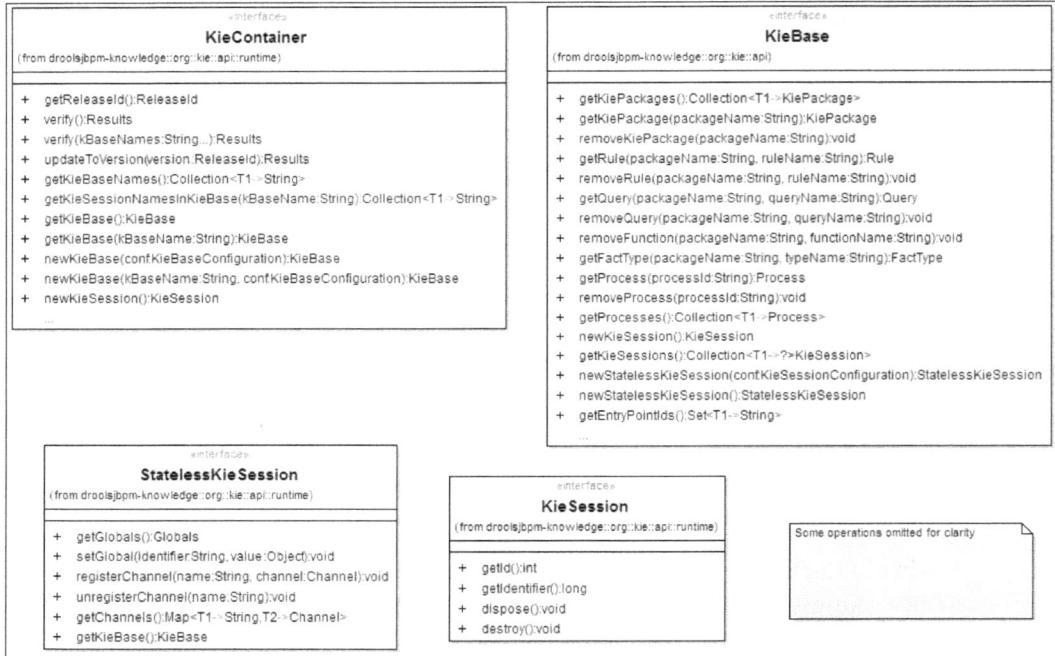

The KIE builder API

It's very likely that you might have already used `KnowledgeBuilderFactory` and `KnowledgeBuilder` to set up `KnowledgeBase`: KnowledgeBuilder parses the knowledge source files (process `bpmn` files, Drools `.drl` rule, and so on), and turns them into KnowledgePackage that KnowledgeBase can use. The resources are identified and added by type (`ResourceType` enum). KnowledgeBase is deprecated, but `KieBase` is actually implemented by `KnowledgeBaseImpl`.

The KIE API gives you tools specialized in managing KieModule: file creation and resource assembling, module dependency management, and building and deployment to Maven repositories. The following class diagram shows the main builder related classes (for a quick reference purpose only).

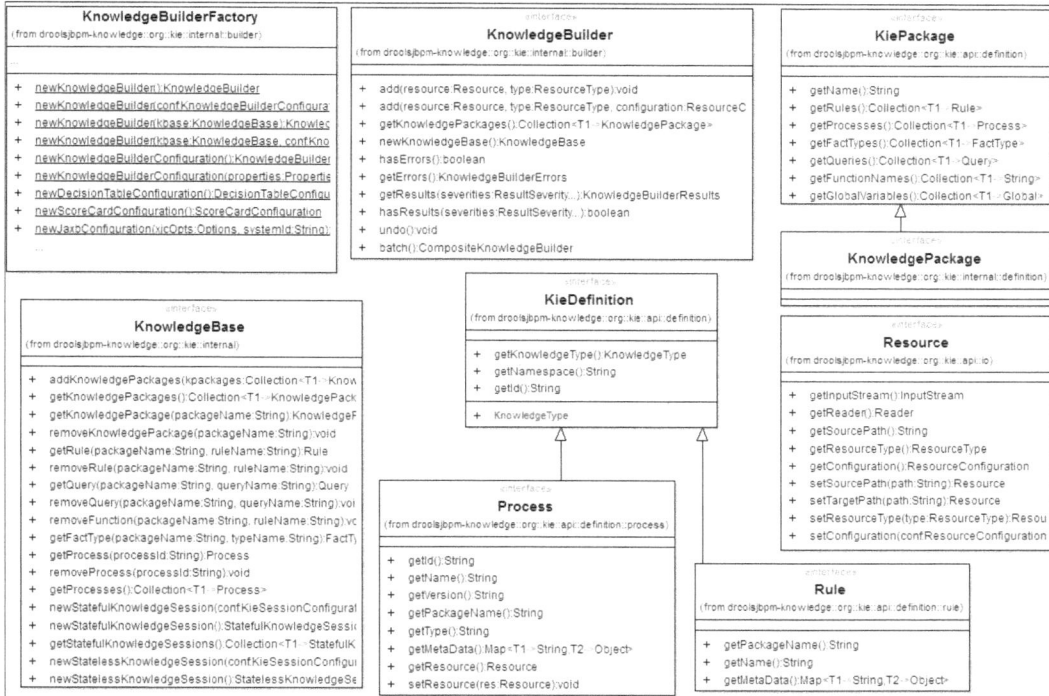

You can find the examples in the `KieBuilderTest` and `KieResourceTest` classes.

KieResources

The resource is a contract representation of a knowledge element (process, rule, and so on) or a resource that indirectly can be used to load a Kie module (for example: a path to `kmodule.xml`).

The `KieResources` factory eases the task of handling objects in the forms of `org.kie.api.io.Resource`; for example:

```
Resource res=ks.getResources().newFileSystemResource (new
File("/jbpm-constructs/target/classes/"));
```

This resource represents the path that contains a KIE module.

KieModule

While KieContainer represents an abstraction, the very nature of KieModule is based on business asset files (resources): KieModule is a container of all the resources needed to define a set of KieBase classes.

> The module project source structure must be compliant with the standard layout for a Maven project (such as `src/main/resources`).

- `pom.xml` defining the KieModule Maven GAV
- `kmodule.xml` declaring KieBase, KieSession, and their properties
- knowledge artifacts

KieModule tracks the module dependencies from other Kie modules and from other plain JAR archives thanks to the `pom.xml` file.

KieBuilder

KieBuilder allows you to build KieModule by adding resources and configuration files through a set of model classes (metamodels), which represent the key KieModule components:

- `KieModuleModel`: A KieModule abstraction
- `KieBaseModel`: A KieBase abstraction
- `KieSessionModel`: A KieSession abstraction

The memory-based filesystem class (`KieFileSystem`) helps you with the creation/writing of the KIE module files (`pom.xml` and `kmodule.xml`). The following class diagram shows `KieBuilder` and the related classes (details ahead).

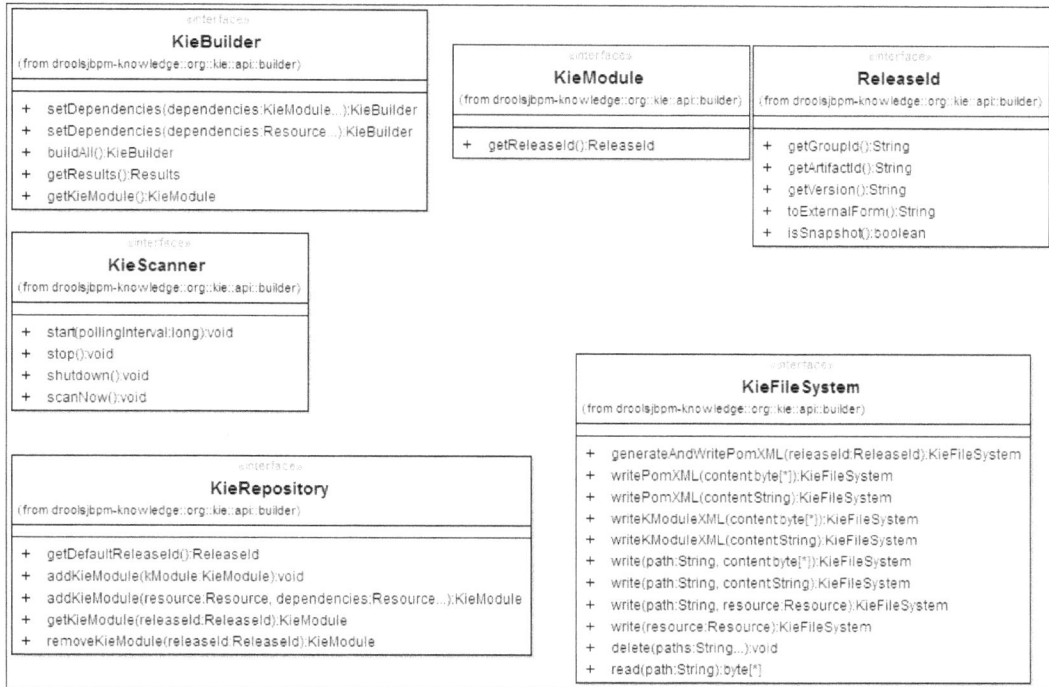

Let us see a practical example of KIE metamodel creation and usage in order to set up and install a KIE module (with dependencies) from scratch into a Maven repository.

You can find the complete example in the `KieBuilderTest` class (`testBuilderWithModels`):

```
KieServices ks = KieServices.Factory.get();
// create the KIE module model
KieModuleModel kmodule = ks.newKieModuleModel();
// create the KieBase model
KieBaseModel kieBaseModel = kmodule.newKieBaseModel("KBase");
// create the KieSession model
KieSessionModel ksession1 =
kieBaseModel.newKieSessionModel("KSession").setDefault(true);
KieFileSystem kfs = ks.newKieFileSystem();
ReleaseId rid = ks.newReleaseId("com.packt.masterjbpm6",
"pizzaDeliveryNew ", "1.0");
// generate pom.xml file
kfs.generateAndWritePomXML(rid);
// and write the <kmodule> xml file
kfs.writeKModuleXML(kmodule.toXML());
```

When your file set is ready, pass `KileFileSystem` (content) to the builder:

```
KieBuilder kieBuilder = ks.newKieBuilder(kfs);

// add dependencies (here, we put jar Files as Resources but you
// can use one or more KieModule too)
Resource dependencyRes = ks.getResources().newFileSystemResource(new
File("c:/temp/pizzadelivery-1.0.jar "));
kieBuilder.setDependencies(dependencyRes);
```

We can now perform the "build." The build compiles all module knowledge packages and Java classes, validates the configuration files (pom.xml and kmodule.xml), and finally, installs the module in the local KIE repository:

```
kieBuilder.buildAll();
if (kieBuilder.getResults().hasMessages(Level.ERROR)) {

}
```

Programmatically creating KieModule means you have to create the object in a file-oriented way, for example:

```
String myprocess= "<?xml version=\"1.0\" encoding=\"UTF-8\"?> \n
<definitions id=\"Definition\"\n" +

kfs.write("src/main/resources/process.bpmn", myprocess);
```

The runtime will create the file following your KieModule filesystem structure.

Repositories and scanners

Maven repositories, as we already pointed out, are an important piece of the new Kie architecture: the repository service allows you to manage module installation and dependency resolution with the internal KIE repository:

```
KieServices ks = KieServices.Factory.get();
KieRepository kr = ks.getRepository();
```

To add a module to the KIE repository store, you must provide the path to the `kmodule.xml` file or the path to the built KIE module JAR file:

```
Resource kieresource= ks.getResources().newFileSystemResource(new
File("c:/Users/simo/git/masterjbpm6/pizzadelivery/target/classes/"));
// or
ks.getResources().newFileSystemResource(new
File("c:/Users/simo/git/masterjbpm6/pizzadelivery/target/
pizzadelivery-1.0.jar"));
```

```
// add to the KIE repo
KieModule kModule = kr.addKieModule(kieresource);
// load and use the module
KieContainer kContainer = ks.newKieContainer(kproj.getReleaseId());
```

The `addKieModule` method accepts the optional module resource dependencies (again, in the form of a `kmodule.xml` path or a path to a JAR archive).

To load a module from the repository is as simple as follows:

```
ReleaseId releaseId = ks.newReleaseId("com.packt.masterjbpm6",
"pizzadelivery", "1.0-SNAPSHOT");
KieModule kModule=kr.getKieModule(releaseId);
```

The repository service wraps the internal Maven KIE repository services as well as the `KieScanner` service that we are now going to see.

KieScanner

KieScanner is a monitor for Maven repositories (both local and remote) used for automatically detecting whether there are updated releases for a given KieModule: in case, a new deployed artifact for the module is found, the scanner updates KieContainer to reflect the changed definitions (KieBase, KieSession, and so on). The KIE module KieBase(s) is rebuilt, and all the new KieSessions created from KieContainer will use the updated KIE module definitions.

The scanner can perform a blocking scan update (the `scanNow()` method), which returns after the (eventual) update process is completed, or a background scanning process (the `start(long pollingInterval)` and `stop()` methods).

```
KieServices ks = KieServices.Factory.get();
ReleaseId releaseId = ks.newReleaseId("com.packt.masterjbpm6",
"pizzadelivery", "1.0-SNAPSHOT");
KieContainer kieContainer = ks.newKieContainer(releaseId);
// bind the scanner to the container
KieScanner scanner = ks.newKieScanner(kieContainer);
// synchronous scanner
scanner.scanNow();
```

> The scanner works only when the paired KieContainer's module Maven version (the V from its GAV) is not a FIXED version. Only modules with a version with the qualifier SNAPSHOT, LATEST, or RELEASE, or ranged versions are processed. See Maven versions reference for additional help.

The scan operation performs the following actions:

- Build the new Kie module, searching for errors
- Update old module dependencies
- Update the old module assets, compile them, and rebuild the module knowledge bases
- If no build error is detected, the updated module is added to the KIE module repository

If there are new or updated classes in use by the knowledge base, then this is fully recreated; otherwise, its resources are incrementally updated. Obsolete knowledge bases and sessions (whose definition has been removed) are deleted from the Kie container. The general advice with the scanner is to be very cautious and to evaluate its impacts on a case-by-case basis.

The example `KieScannerTest` class provides you with two test methods:

- `testScannerUpdateNewSession`: This verifies whether a Kie module process definition gets updated after the scan process by creating a new session and verifying that the update process variable returns a different value from the original definition
- `testScannerSameSessionAfterUpdate`: This verifies whether after the scan, the existing session continues using its old process definition, while a new KIE session picks up the updated process definition

The scanner is a nice improvement over the previous jBPM knowledge base update mechanism (KnowledgeAgent) since it works in tight integration with Maven and provides the implementer with an asset-oriented programming style in handling KIE modules, Kie project sources, and assets. This great addition makes jBPM fit a lot better in the typical agile, lean development environment. Just think about the possibilities you have when integrating with **Continuous Integration (CI)**, deployment, and automated test tools. You may have a scanner process that checks your nightly build Maven repo trunk, updates your KIE module with the latest development version of your assets, and triggers your automated test suite.

> The KieScanner API implementation and utility classes belong to the `kie-ci` project (`https://github.com/droolsjbpm/drools/tree/master/kie-ci`).

The Scanner API also provides a Maven helper class, which manages artifact lookup and deployment to the system Maven repository:

```
MavenRepository repo = MavenRepository.getMavenRepository();
List<DependencyDescriptor> dependencies = repo.
getArtifactDependecies("com.packt.masterjbpm6:pizzadelivery:1.0");
Artifact module = repo.resolveArtifact(ks.newReleaseId(
   "com.packt.masterjbpm6", "pizzadelivery", "1.0"));
```

The `KieScannerTest` jUnit test class exercises the scanner and the builder API. It creates and deploys a new release for the `pizzadelivery` KieModule (the `buildModuleForScannerUpdate` method) and then, starts the scanner update process (the `testScannerUpdate()` method).

KieLoggers

The KieLoggers factory allows you to create audit loggers that produce log traces of all the events occurring during the execution of a specific KIE session. The following types of loggers are available:

- **File based logger**: Logger to file with a default `.log` extension; it traces the event in an XML serialized format:

  ```
  KieRuntimeLogger logger = loggers.newFileLogger(ksession,
          "c:/temp/kielogger");
  ```

  See the `KieLoggersTest.testLoggers` method for the complete example.

- **Console based logger**: Traces the log to the standard output:

  ```
  KieRuntimeLogger logger = loggers.newConsoleLogger(ksession);
  ```

  Here you have an example when running the `testRuleWithConsoleLogger` method; you can see the insertion of the Drools fact and the Drool rule triggering:

  ```
  13:31:03.459 [main] INFO  o.d.c.a.WorkingMemoryConsoleLogger
  - OBJECT ASSERTED
  value:com.packt.masterjbpm6.pizza.model.Order@12e13d86
  factId: 1


  13:31:03.708 [main] INFO  o.d.c.a.WorkingMemoryConsoleLogger -
  BEFORE RULEFLOW GROUP ACTIVATED
  group:masterRuleGroup[size=1]
  ```

```
13:31:03.724 [main] INFO  o.d.c.a.WorkingMemoryConsoleLogger
- BEFORE ACTIVATION FIRED rule:checkorder
activationId:checkorder [1] declarations:
$o=com.packt.masterjbpm6.pizza.model.Order@12e13d86
ruleflow-group: masterRuleGroup
```

- Threaded logger: Same as the file-based logger but executes writes to the file in an asynchronous fashion; it features an option to set the write (flush) interval period in milliseconds, for example:

```
// update the log file every 5 seconds
KieRuntimeLogger logger =
loggers.newThreadedFileLogger(ksession,
        "c:/temp/kie_threaded", 5000);
```

See the `testRuleWithThreadedLogger` example for the complete example.

Logger classes extend `WorkingMemoryLogger`, which implements all the available event listener interfaces: `Process`, `Agenda`, `Rule`, and (KIE) KnowledgeBase. Since several events are generated, you're given the ability to control event filtering with the following methods: `addFilter`, `removeFilter`, and passing an `ILogEventFilter` implementing class. We can declare and configure the KieSession loggers directly in the `kmodule.xml` file, for example:

```
<ksession

<fileLogger id="filelogger" file="mysession.log" threaded="true"
interval="10" />
<consoleLogger id="consolelog" />
</ksession>
```

The following class diagram shows the loggers and the event listener interfaces:

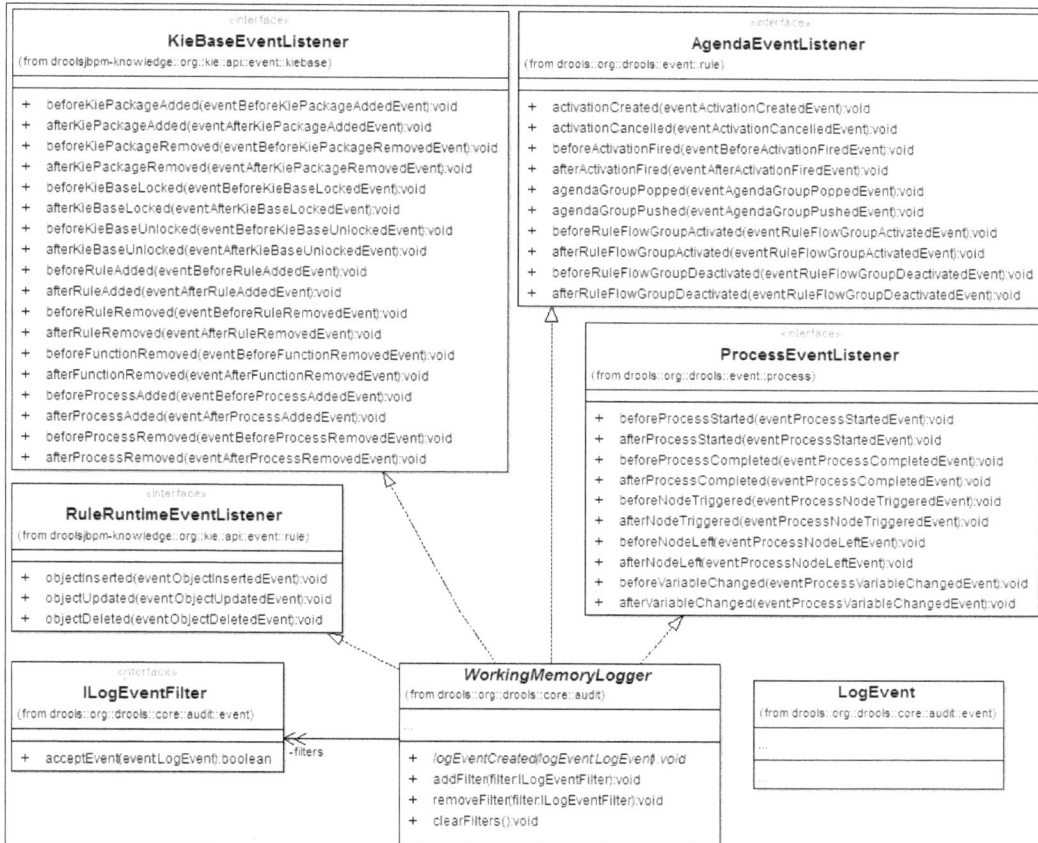

KieStoreServices

`org.kie.api.persistence.jpa.KieStoreServices` is an interface that defines the contract for the KIE session persistence service. Its default implementation is the `org.drools.persistence.jpa.KnowledgeStoreServiceImpl` class. Let us have a look at how `KieStoreServices` can be used to restore KIE sessions by using the session ID. The following example is an excerpt of the `StorageTest.testRuleWithStorageServer` method. It demonstrates how you can load your Kie session safely from the persistence store and execute the business process consistently.

```
KieServices ks = KieServices.Factory.get();
KieStoreServices storeservice = ks.getStoreServices();
KieContainer kContainer = ks.getKieClasspathContainer();
```

```
KieBase kbase = kContainer.getKieBase("kbase");
// initialize the Session Environment with EMF and the TM
Environment env = EnvironmentFactory.newEnvironment();
    env.set(EnvironmentName.ENTITY_MANAGER_FACTORY, super.getEmf());

// current Bitronix transaction manager
  env.set(EnvironmentName.TRANSACTION_MANAGER,
    TransactionManagerServices.getTransactionManager());
// create the session
  ksession = storeservice.newKieSession(kbase, null, env);
// perform the Rule task test
  testRule();
  long id = ksession.getIdentifier();
// dispose the session
  ksession.dispose();
// reload the session given its ID
  KieSession loadedsession = storeservice.loadKieSession(id, kbase,
null,env);
// check it is the same session
  assertEquals(id, loadedsession.getIdentifier());
// rerun the test on the loaded session
ksession = loadedsession;
  testRule();
```

RuntimeManager service and the engine

RuntimeManager has been introduced to simplify the configuration of KieBase (KnowledgeBase) and KieSession (KnowledgeSession). Its main duty is to manage and create instances of RuntimeEngine according to predefined strategies (see the *Runtime strategy* section).

org.kie.api.runtime.manager.RuntimeEngine is the entry point to the engine services; its main purpose is to provide the user with pre-configured and ready-to-use engine components:

- KieSession
- TaskService
- AuditService

`org.kie.api.runtime.manager.RuntimeManager` unburdens the user from the development of a typical boilerplate code; it sets up the execution environment for processes (wrapping the KieSession and the task service into the RuntimeEngine) and manages the following Drools services:

- **Scheduler service**: The scheduler service manages timer-based jobs for execution (we have seen it in *Chapter 4, Operation Management*, and *Chapter 5, BPMN Constructs* with *Async task*)

- **TimerService**: Implements timer services for the sessions

The runtime manager registers the following items on the session:

- Process workitem handlers (also the default `human task` workitem handler which is responsible for managing task nodes)

- Global variables

- Event listeners

Runtime strategy

`RuntimeManager` implements a runtime strategy that lets you choose how to manage your KieSession life cycle; let us see the available strategies:

- **Singleton (default jBPM strategy for a Kie module)**: The runtime manages only one shared `RuntimeEngine` instance (only one Kie session is active and shared)

- **PerProcessInstance**: The manager uses a dedicated Kie session for each process instance; the Kie session life cycle spans the process instance duration

- **PerRequest**: Invoking the `getRuntimeEngine()` method returns a new `RuntimeEngine` instance (creating a new Kie session and task service each time)

`RuntimeManager` must be created from `RuntimeManagerFactory`, calling one of its specialized factory methods according to the chosen runtime strategy (`newSingletonRuntimeManager()`, `newPerRequestRuntimeManager()`, or `newPerProcessInstanceRuntimeManager()`) and passing an instance of `org.kie.api.runtime.manager.RuntimeEnvironment`.

Choosing the right strategy

You shall choose the right runtime strategy mainly depending on your business and system specifications. Requirements may constrain you to keep isolated jBPM session working memories (for example, one session per process instance); in other words, each session owns its rules, facts, and objects. This could be the case for short-lived processes in a heavily concurrent system where you need low resource contention and high throughput.

The singleton strategy, on the other hand, manages a single-thread safe session (with synchronized access). This could lead to performance issues in highly concurrent environments (web) but would also allow for all jBPM sharing capabilities (facts and globals shared among all processes, scope of signals across your entire working memory, and so on). These are just general insights into the matter, and you should aim at evaluating your own strategy pros and cons according to your system and functional requirements.

The RuntimeEnvironment class

This class encapsulates the environment configuration required by `RuntimeManager`; we instantiate it by using the `RuntimeEnvironmentBuilder` helper class:

```
// preconfigured environment with disabled persistence
RuntimeEnvironmentBuilder
builder=RuntimeEnvironmentBuilder.Factory.get()
      .newDefaultInMemoryBuilder();
// with enabled persistence: emf is your EntityManagerFactory
RuntimeEnvironmentBuilder.Factory.get().newDefaultBuilder().entityMan
agerFactory(emf).persistence(true);
```

`RuntimeEnvironmentBuilderFactory` has several helper methods targeted to create preconfigured specialized `RuntimeEnvironmentBuilder` enabled for persistence, based on the classpath KIE container (`kmodule.xml`), based on a KIE module (JAR file), and so on. Since the builder wraps all the configuration details, it exposes methods to perform the following:

- Add an asset (a BPMN process, Drools rule, and so on)
- Set custom `UsergroupCallback`
- Set a Knowledge Base (in case your `RuntimeEnvironmentBuilder` is not a KIE module classpath builder)

- Set Entity Manager Factory to enable JPA persistence

```
RuntimeManagerFactory managerFactory =
RuntimeManagerFactory.Factory.get();

// pass the RuntimeEnvironment we get from the EnvironmentBuilder
RuntimeManager rtManager=managerFactory.newSingletonRuntimeManager
(builder.get());
```

RuntimeManagers are identified by unique identifiers. The runtime won't accept the creation of RuntimeManager with the same id of another active RuntimeManager. The following diagram shows the interactions that take place during the runtime initialization and that involve the main KIE components:

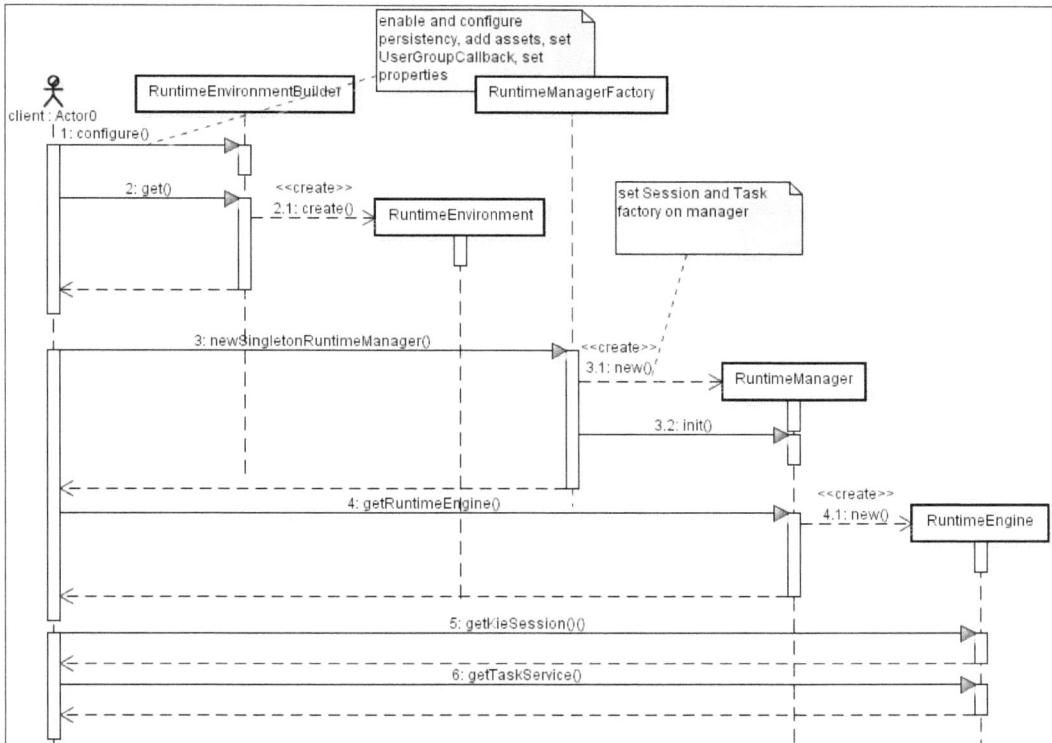

Runtime Context

RuntimeManager can handle contextual information to look up a specific RuntimeEngine implementation, depending on the chosen strategy; the context is passed as a org.kie.api.runtime.manager.Context generic interface implementation:

- EmptyContext: Context used with Singleton or PerRequest RuntimeManager; no specific information is used

- CorrelationKeyContext: Used with PerProcessInstance RuntimeManager to find RuntimeEngine by using a process instance correlation key

- ProcessInstanceIdContext: Used with PerProcessInstance RuntimeManager to find RuntimeEngine (and the Kie session) by using a process instance ID

```
RuntimeEngine engine = rtManager.getRuntimeEngine(EmptyContext.
get());
// we can now get the initialized services
KieSession ksession = engine.getKieSession();
TaskService taskService = engine.getTaskService();
```

To avoid resource leakage, it is strongly recommended to dispose RuntimeManager at the end of a work session, for example:

```
rtManager.close();
```

KieSession

KieSession is a stateful session that maintains its conversational state with the engine, across multiple interactions. It is the best way to interact with the engine. Sessions are created starting from KieContainer, KieBase, or configured RuntimeEngine, which always delegates to KieBase but gives you the flexibility of choosing a session runtime policy.

Depending on the enabling of persistence, the session is created as follows:

- **In-memory session**: All data related to session and the engine status is kept in memory and lost on engine restart

- **JTA session**: Session persisted through a JPA EntityManager and a JTA transaction manager

To create a new (stateful) KieSession, we configure the environment and we use
`JPAKnowledgeService`:

```
Environment env = KnowledgeBaseFactory.newEnvironment();
EntityManagerFactory emf= Persistence.createEntityManagerFactory(
"com.packt.masterjbpm6.persistenceunit" ));
env.set( EnvironmentName.ENTITY_MANAGER_FACTORY,emf);
env.set( EnvironmentName.TRANSACTION_MANAGER,
bitronix.tm.TransactionManagerServices.getTransactionManager());

StatefulKnowledgeSession ksession =
JPAKnowledgeService.newKieSession( kbase, null, env );
```

The example uses **Bitronix Transaction Manager** (BTM) (more on this in the
Persistence and transaction section).

The returned `StatefulKnowledgeSession` is of type
`CommandBaseStatefulKnowledgeSession`, which decorates the session
implementation (`StatefulKnowldgeSessionImpl`) with a command service
of type `SingleSessionCommandService` (see the following class diagram).

`SingleSessionCommandService` transparently manages, through the
`TransactionInterceptor` class, the persistence of the stateful session
by the JPA `SessionInfo` entity class.

Stateless KieSession

The stateless KIE session is a wrapper to a stateful session, which the runtime creates
and disposes for the duration of a single command execution so that it does not
maintain a conversational state and cannot persist.

Globals

KieSession manages `globals`; globals are objects used to pass information into the
engine that can be used in processes or rules. Globals are shared across all processes
and rule instances. Let us see what KieSession methods can handle them:

* `getGlobals()`: Returns the internal globals resolver
* `getGlobal (String)`: Returns the global object given its identifier
* `setGlobal(String, Object)`: Sets the global object assigning it an identifier

The following class diagram shows details for the Session classes:

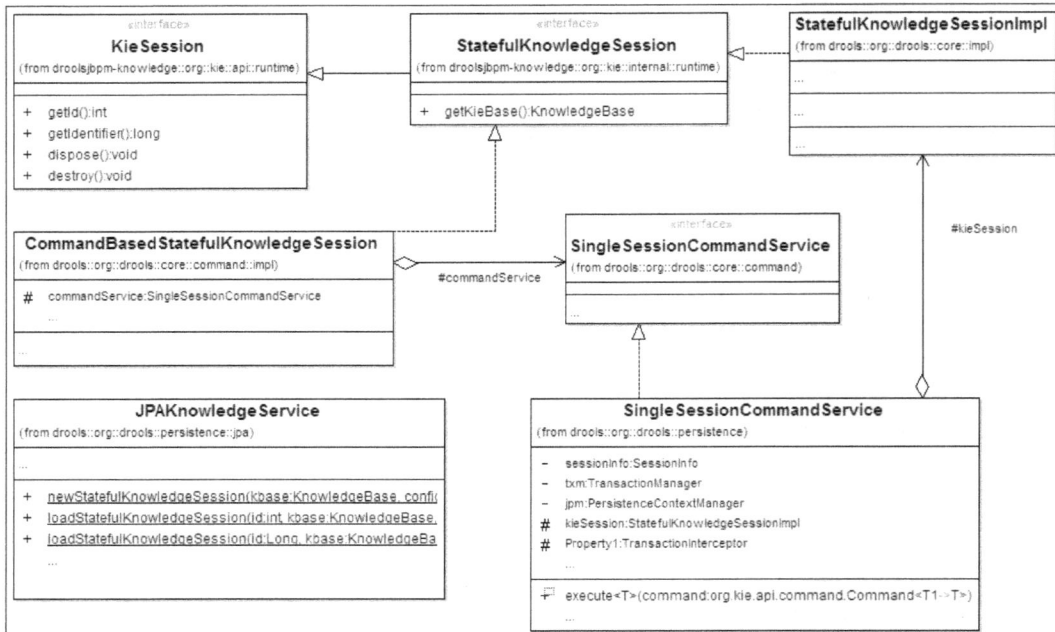

The CommandExecutor interface

All KIE sessions (both stateless and stateful) implement the `CommandExecutor` interface, a service that lets you execute a single command or a batch of commands. Let us look at some of them:

- **Process/Task**: `CompleteWorkItemCommand`, `AbortWorkItemCommand`, `StartProcessCommand`, and so on

- **Drools Rules**: `FireAllRulesCommand`, `GetFactHandleCommand`, and so on

- **Runtime**: `GetGlobalCommand`, `SetGlobalCommand`, and so on

Normally, you would call the higher-level jBPM API methods (using the session or the task service reference), but sometimes, it could be handy to use the command classes for batching, scheduling purposes, or management operations. All the command classes support XML binding thanks to standard annotations (`@XmlRootElement`, `@XmlAttribute`, `@XmlElement`, and so on) and so, can be easily serialized or passed remotely. Commands can be created using `CommandFactory` or by simply instantiating a specific command class, which is then executed by invoking the session `execute` method.

Let us now have a look at how these commands can be created and executed from the following code:

```
// create the command and execute
Command startProcess= CommandFactory.newStartProcess
("pizzadelivery");
ksession.execute(startProcess);
```

Alternatively, you can instantiate the command on your own:

```
GetProcessInstanceCommand getProcessInstanceCommand = new
GetProcessInstanceCommand();
getProcessInstanceCommand.setProcessInstanceId(
processInstance.getId());
ProcessInstance processInstance = ksession.execute(
getProcessInstanceCommand );
```

Batch execution

The session supports the execution of a batch of commands by a specialized `BatchExecutionCommand`. Here, we rewrite the `RuleTaskTest` test class of *Chapter 5, BPMP Constructs*, by using three different commands (see the `CommandsTaskTest. testRuleWithCommand` method):

```
StringBuffer orderdesc = new StringBuffer();
List<Command> batchcmds = new ArrayList<Command>();
batchcmds.add(CommandFactory.newSetGlobal("newnote", orderdesc));
Order order = new Order();
order.setCost(200);
batchcmds.add(CommandFactory.newInsert(order));
batchcmds.add(CommandFactory.newSetGlobal("orderglobal", order));
batchcmds.add(CommandFactory.newStartProcess("rule"));
ExecutionResults results = ksession.execute(CommandFactory
        .newBatchExecution(batchcmds));
```

`BatchExecutionCommand` executes the enlisted commands in the exact order that they have been added:

```
List<Command> pizzabatchcmds = new ArrayList<Command>();
pizzabatchcmds.add(CommandFactory.newStartProcess("pizzadelivery"));
pizzabatchcmds.add(CommandFactory.newStartProcess("pizzadelivery"));
ksession.execute(CommandFactory.newBatchExecution(pizzabatchcmds));
```

Note that `BatchExecutionCommand`, like all Command classes, supports serialization, so you can pass it remotely for execution or easily persist it for scheduled processing.

> Even `CompositeCommand` can execute multiple commands sequentially, but it supports human task commands only (inheriting from `TaskCommand`). This command is used internally by the task service (see the *Human Task service* section).

Event listeners

KIE session can register multiple specialized listeners for different kinds of event notifications:

- **Process**: `ProcessEventListener` is related to process instance execution (we saw ProcessEventListener in the *KieLoggers* section).
- **Rule**: `RuleRuntimeEventListener` for events related to facts.
- **Agenda**: `AgendaEventListener` for events related to Drools Agenda. Agenda is a Drools component that keeps track of rule activations (rule matching) and rule action execution.

By providing a custom implementation of one of these interfaces, you can get the filtered session events.

> All the logger classes obtained from the `KieLoggers` service extend the abstract `WorkingMemoryLogger` class, which implements the preceding three interfaces. We will see more on event and auditing in the *Audit and history logs* section.

We can declaratively register listeners inside the `kmodule.xml` file:

```
<ksession name="ksession">
<listeners>
<ruleRuntimeEventListener
type="com.packt.masterjbpm6.RuntimeEventlistener" />
<agendaEventListener type="com.packt.masterjbpm6.AgendaEventlistener"
/>
<processEventListener type="com.packt.masterjbpm6.
ProcessEventlistener" />
</listeners>
</ksession>
```

Channels

Channels are communication hoses that can be established between your jBPM application and the engine, through your KIE working session. Their main purpose is to allow sending objects from the session working memory to a generic external process or function.

Let us see the basics: You must implement the `org.kie.api.runtime.Channel` interface and register it with the session, assigning a name to the channel:

```
public class RulesAppChannel implements Channel {
    // handle the channel object sent
public void send(Object object) {

    }
}
ksession.registerChannel("appChannel", new RulesAppChannel());
```

The channel can be used to notify the jBPM application, for instance, of a rule execution as follows:

```
when
        $o: com.packt.masterjbpm6.pizza.model.Order (cost>100)
    then
    {

    channels["appChannel"].send("urgent!");
    }
```

The list of existing channels can be retrieved from the session by calling:

```
Map <String, Channel> channels=ksession.getChannels();
```

Check the `ChannelsTaskTest.testRuleWithChannel()` method for a working example.

Human Task service

We introduced the human tasks in the previous chapter; we described the human task state transitions, task rule assignments, and task operations performed by the stakeholders.

You can get the `org.kie.api.task.TaskService` task service from `RuntimeEngine` (the `getTaskService()` method), which is a wrapper for the service; the task service is created and wrapped into `RuntimeEngine` (and the underlying session) by `RuntimeManager`, which uses `TakServiceFactory`:

```
InternalTaskService internalTaskService = (InternalTaskService)
taskServiceFactory.newTaskService();
```

The factory used to instantiate `TaskService` is as follows:

- `LocalTaskServiceFactory`: To be used in non-CDI environments

The factory configures the task service with the following references:

- `EntityManagerFactory` instance (taken from `RuntimeEnvironment`).
- `UserGroupCallback` (taken from `RuntimeEnvironment`). If no custom `UserGroupCallback` implementation is provided, a default `MvelUserGroupCallbackImpl` is used; this loads the `UserGroupsAssignmentsOne.mvel` file from `jbpm-human-task-core-6.2.0.Final.jar` (the `org\jbpm\services\task\identity` package).
- Register task event listeners (instances of the `TaskLifeCycleEventListener` interface).
- `TaskDeadlinesService`: This service is responsible for the management of the deadlines defined for a task and for triggering notifications (we talked about escalations and notifications in *Chapter 5, BPMN Constructs*).

The factory also shares, at the environmental level, the following:

- Default `DefaultUserInfo` instance (loaded with data from a `userinfo.properties` file in the classpath, if any)

The task service leverages Commands to execute all common task operations; commands are executed via `CommandService`. Let us look at this service class and its inner working details.

CommandService

RuntimeManager creates two different types of TaskService:

- `CommandBasedTaskService`: The default task service implementation
- `SynchronizedTaskService`: A synchronized task service instance that wraps `CommandBasedTaskService`. It is created using the Singleton runtime strategy

`CommandBaseTaskService` delegates the API executions to its internal `CommandService` executor. This is CommandExecutor and performs all the task API calls as in the KIE session (see the dedicated *The CommandExecutor interface* section), for example:

```
public void start(long taskId, String userId) {
  executor.execute(new CompositeCommand<Void>(
    new StartTaskCommand(taskId, userId),
    new CancelDeadlineCommand(taskId, true, false)));
}
```

The start (task) method, for instance, is executed as `CompositeCommand` built with two TaskCommand specializations:

- `StartTaskCommand`: It performs the task start operation by changing the state, triggering events on listeners, and so on
- `CancelDeadlineCommand`: Cancel the matching deadline (if any) for this task's Start event (see the *Deadlines and escalations* and the *Notifications* sections in *Chapter 5, BPMN*)

`CompositeCommand` first executes the command from the varying argument `commands` and then, the `mainCommand` command. Its constructor signature is as follows:

```
public CompositeCommand(TaskCommand<T> mainCommand,
TaskCommand<?>...commands) {

}
```

TaskCommand and UserGroupCallback

All task commands inherit from the `UserGroupCallbackTaskCommand` class; they call specific parent class methods on their `execute` method in order to trigger updates to the `ORGANIZATIONALENTITY` database table.

The TaskContext command

Upon instantiation, each task command is given `TaskContext` from `CommandExecutor`; the context duties are as follows:

- Provides a `JPATaskPersistenceContext` instance, which handles all the database-related operations
- Triggers task lifecycle-related events
- Provides the core task-related services to the command

The main task services provided are as follows:

- **Instance service** (TaskInstanceService): The core service that implements the WS Human Task specification with respect to the task life cycle

- **Query service** (TaskQueryService): It returns read-only task instance information such as a list of tasks assigned to a user, potential owners for a given task, and so on

- **Content service** (TaskContentService): It manages task content data (input and output)

- **Deadline service** (TaskDeadlineService): It manages the scheduling of task deadlines

- **Attachment service** (TaskAttachmentService): It deals with task attachment management

- **Admin service** (TaskAdminService): It provides out-of-the-standard task life cycle operations such as task removal and task archival

Tasks can be removed (literally deleted from the jBPM persistent store, making them unavailable for user assignments) but only after they have been marked as "archived."

Transactions and interceptors

TaskService CommandService is implemented by a TaskCommandExecutorImpl class, which, during initialization, is decorated with an org.jbpm.services.task. persistence.TaskTransactionInterceptor interceptor. The interceptor wraps each execute method invocation of the command service between transaction boundaries managed by the Drools JtaTransactionManager.

Notification service

In *Chapter 5, BPMN Constructs*, we talked about the human task escalations and notifications features. The jBPM notification service relies on email messaging; thus, to work successfully, it requires the following:

- A mail session properly configured

- A UserInfo class instance defined; this provides the user's e-mail address to notify

jBPM loads the mail session by a **Java Naming and Directory Interface (JNDI)** lookup; you can set the JNDI name through the org.kie.mail.session system property or, alternatively, provide the JNDI name mail/jbpmMailSession in your application server.

If no JNDI resource is found, jBPM fallbacks to a plain `email.properties` file loaded from the classpath.

The file defines the following properties (example values are provided):

```
mail.smtp.host=localhost
mail.smtp.port=2345
mail.from=alerts@packt.com
mail.replyto=replyTo@packt.com
```

We covered the `UserInfo` class in *Chapter 5*, *BPMN Constructs*.

The TaskFluent class

The `TaskFluent` class is a helper class that lets you conveniently configure and create a new `Task` instance and add it to the persistent store through the task service (see the `FluentTest` test class):

```
TaskFluent fluent = new TaskFluent();
fluent.setName("adhoc Human Task");
fluent.addPotentialUser("Luigi").setAdminUser("Administrator");
Task task = fluent.getTask();
// add the task
long taskid = taskService.addTask(task, new HashMap<String,
Object>());
```

> The WS-HumanTask specification defines the role of the administrator as the one who can manage the life cycle of the task, even though he might not be enlisted among the task potential owners. With releases older than jBPM 6, an "Administrator" user was created by default by jBPM.

The `addTask` operation requires you to add at least one potential business administrator for the task (by the `setAdminUser()` or `setAdminGroup()` method). The business administrator entity (user or group) is verified by the runtime with the current `UserGroupCallback` to check whether it exists. To check whether business administrators are task potential owners search the `PEOPLEASSIGNMENTS_BAS` jBPM database table.

To ease the test configuration, all our jUnit test classes are configured with a custom user callback, which lets all users pass (see the `MyUserCallback` class), so that "Administrator," "boss," or whatever makes no difference.

The runtime will evaluate the task assignments and their deadlines, and will store the task data; the engine assigns to the task, the initial state CREATED.

The FluentTest example shows how it is possible to programmatically create, add, assign, start, and complete new tasks outside the scope of a process definition (ad hoc tasks).

TaskFluent lets you attach a new task to a given process instance. Since the new task has no incoming/outgoing connections, this could be of limited use, but it fits very well with ad hoc processes (see *Chapter 5, BPMN Constructs*), for example:

```
ftask.setWorkItemId("default-singleton");
ftask.setProcessId(PROCESS_ID);
ftask.setProcessInstanceId(processinstance.getId());
ftask.setProcessSessionId(ksession.getId());
```

Audit and history logs

Auditing is the ability of the engine to let the user configure the collection and the retrieval of events relative to the runtime process execution. We introduced auditing and BAM in *Chapter 4, Operation Management*, and now, we are going to see how to leverage the engine services to implement auditing at various levels.

When persistence is enabled, ready-to-use preconfigured AuditService (implemented by JPAAuditLogService) can be borrowed from RuntimeEngine; it returns process, node, and variable runtime audit data (see the class diagram) from the three standard jBPM auditing database tables (ProcessInstanceLog, NodeInstanceLog, and VariableInstanceLog). The audit service stamps each log entry with an OwnerID attribute, which matches the unique identifier of its RuntimeManager.

```
AuditLogService auditservice=engine.getAuditLogService();
List<ProcessInstanceLog> pizzadeliveryLogs= auditservice.findActivePro
cessInstances("pizzadelivery");
```

This AuditService usage approach (basically as a query service) is a solution that can be good for BAM or history analysis; however, if we want to collect audit data in real-time, we have to register a listener with the KIE session (see the *Event listeners* section or the next section).

AuditLoggerFactory

This factory can create ready-to-use JPA- or JMS-enabled audit loggers (see the following class diagram):

- **JPA**: A synchronous logger that by default persists the audit events between engine JTA transaction boundaries
- **JMS**: Asynchronous logger aimed at JMS queue integration

Remember that the logger has to be bound to KieSession to start logging:

```
ksession.addEventLister(listener);
```

We can set a different EntityManager on the JPA logger by providing custom EntityManagerFactory to a directly instantiated (no factory) JPAAuditLogService and, optionally, choosing an auditing strategy:

```
EntityManagerFactory emf =
Persistence.createEntityManagerFactory("com.packt.masterjbpm6.persist
enceunitApp");
AuditLogService logService = new JPAAuditLogService(emf,
PersistenceStrategyType.STANDALONE_LOCAL);
```

This allows us to adapt the engine logging service to our persistence layer configuration and transaction management: local entity manager and JTA. The main purpose of the strategy is to instruct jBPM to manage transactions when auditing in different application environments.

The KIE auditing strategies are as follows:

- **KIE_SESSION**: Select this strategy when you use the entity manager only with KieSession (default behavior)
- **STANDALONE_LOCAL**: Choose this strategy when your application uses the entity manager outside the scope of KieSession
- **STANDALONE_JTA**: Same as the previous strategy but uses `java:comp/UserTransaction` or `java:jboss/UserTransaction` to resolve the transaction (application servers only)

We show the relevant classes in the following class diagram:

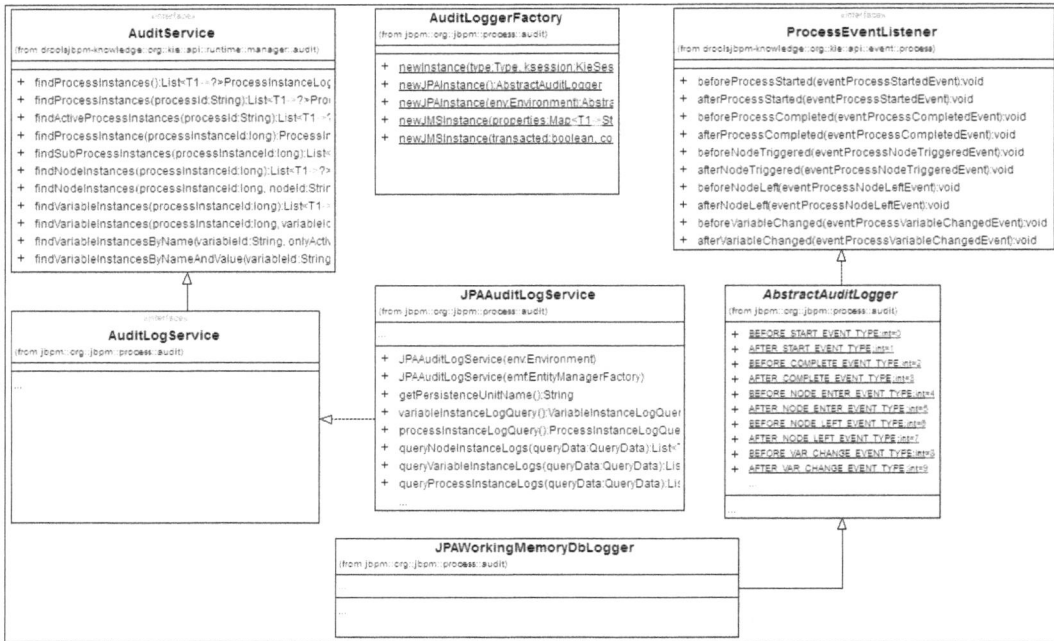

Custom logger

To implement custom logging of audit events, you have to extend the `AbstractAuditLogger` class or provide a new implementation for the `ProcessEventListener` interface.

The supported audit events are as follows:

```
Process:
    BEFORE_START_EVENT_TYPE = 0;
    AFTER_START_EVENT_TYPE = 1;
    BEFORE_COMPLETE_EVENT_TYPE = 2;
    AFTER_COMPLETE_EVENT_TYPE = 3;
Nodes:
    BEFORE_NODE_ENTER_EVENT_TYPE = 4;
    AFTER_NODE_ENTER_EVENT_TYPE = 5;
    BEFORE_NODE_LEFT_EVENT_TYPE = 6;
    AFTER_NODE_LEFT_EVENT_TYPE = 7;
Variables:
    BEFORE_VAR_CHANGE_EVENT_TYPE = 8;
    AFTER_VAR_CHANGE_EVENT_TYPE = 9;
```

Events are delivered as the following `ProcessEvent` implementation classes: `ProcessNodeEvent`, `ProcessStartedEvent`, `ProcessCompletedEvent`, and `ProcessVariableChangedEvent`. In order to convert the incoming typed events to Log JPA entity classes (that is, `VariableInstanceLog`), you can use a helper class (`DefaultAuditEventBuilderImpl`):

```
AuditEventBuilder builder = new DefaultAuditEventBuilderImpl();
VariableInstanceLog variablelog = (VariableInstanceLog) builder.
buildEvent(event);
// get process variable properties: processId, var. name and value
String processId= variablelog.getProcessId();
String variableId=variablelog.getVariableId();
String value= variablelog.getValue();
// persist
em.persist (variablelog);
```

Persistence and transactions

The default persistence mechanism of the jBPM engine is based on the JPA 2/ Hibernate implementation. Each engine operation (start process, start task, complete task, and so on) is run inside the scope of a transaction. `TransactionInterceptor` demarcates each command execution and eventually, depending on the transaction management used (**Container Managed Transactions (CMT)** or UserTransaction **Bean Managed Transactions (BMT)**), enlists the EntityManager engine in the ongoing transaction. We have seen how both session and task persistence works through `CommandService` and the interceptor architecture.

The default engine persistence configuration boils down to the engine persistence unit (defined in a `persistence.xml` file configuration) and, usually, to a JTA datasource definition at the application server level. jBPM imposes no constraints on the number of entity managers defined; you can obviously have a number of persistence units defined in your application and make multiple entity managers coexist with jBPM Entity Manager. You can have a single database (single Entity Manager) for both jBPM and your application persistence layer or a dedicated jBPM database (and Entity Manager) and make the engine participate in your business application transactions.

> *Chapter 8, Integrating jBPM with Enterprise Architecture, of the jBPM 6.2 user guide explains the detailed configuration and usage of transactions.*

Local transactions with Bitronix

We are going to see a complete jBPM persistence example configuration using local transactions. Bitronix is an open source Transaction Manager framework; it supports JTA API and the **Extended Architecture (XA)** protocol and perfectly fits all cases where a straightforward persistence configuration is needed. The required configuration steps are as follows:

1. Create the datasource (pooled): The datasource will be bound to the jdbc/localjbpm-ds JNDI name as follows:

```
PoolingDataSource pds = new PoolingDataSource();
pds.setUniqueName("jdbc/localjbpm-ds");
pds.setClassName(LrcXADataSource.class.getName());
pds.setMaxPoolSize(5);
pds.setAllowLocalTransactions(true);
pds.getDriverProperties().put("user","sa");
pds.getDriverProperties().put("password","");
pds.getDriverProperties().put("url","jdbc:h2:tcp://localhost
/~/jbpm-db;MVCC=TRUE");
pds.getDriverProperties().put("driverClassName","org.h2.Driv
er");
pds.init();
```

2. Create the jndi.properties file in your classpath resources, which includes the following code:

```
java.naming.factory.initial=bitronix.tm.jndi.
BitronixInitialContextFactory
```

This lets Bitronix context factory initialize the environment and bind transaction service objects to default JNDI names, notably the following:

   ○ User transaction manager at java:comp/UserTransaction
   ○ Tx synchronization registry at java:comp/
     TransactionSynchronizationRegistry

3. Edit your persistence.xml file, specifying the Bitronix datasource name as follows:

```
<persistence-unit name="localjbpm-persistenceunit"
    transaction-type="JTA">
<provider>org.hibernate.ejb.HibernatePersistence</provider>
<!-- match the bitronix datasource uniqueName -->
<jta-data-source>jdbc/localjbpm-ds</jta-data-source>
```

4. Now, you can create your `EntityManagerFactory` as follows:

```
EntityManagerFactory emf
=Persistence.createEntityManagerFactory("localjbpm-
persistenceunit");
```

Managing transactions

jBPM provides an out-of-the-box transactional service to an enterprise application, which is able to participate in the existing transaction right from the calling application, so in case of an error (for example, a custom workitem handler throws an exception or a process node script fails), the engine transaction is marked for rollback and the exception is sent to the caller.

Let us now see a common example of **Entity Manager** (**EM**) and jBPM (managed by Bitronix) working together (please refer to the `AuditTxTest` test class):

```
AuditEntity audit = new AuditEntity();
audit.setDesc("startAudit1");
UserTransaction ut;
try {
  ut = (UserTransaction) new InitialContext()
      .lookup("java:comp/UserTransaction");
  ut.begin();
  em.joinTransaction();
  em.persist(audit);
// start process
  ProcessInstance pi = ksession.startProcess("auditTxProcess");
// new application database insert
  AuditEntity auditproc = new AuditEntity();
  auditproc.setDesc("Audit1:process started");
  em.persist(auditproc);
// commit both process instance and audit entity
  ut.commit();
```

Locking

The default JPA transaction locking scheme used is optimistic. If you need to switch to a pessimistic locking mode, set the following parameter to TRUE in your `org.kie.api.runtime.Environment` instance:

```
EnvironmentName.USE_PESSIMISTIC_LOCKING
```

This forces the engine to hold the lock on an entity (locking of type `LockModeType.PESSIMISTIC_FORCE_INCREMENT`) to ensure that the object is not modified.

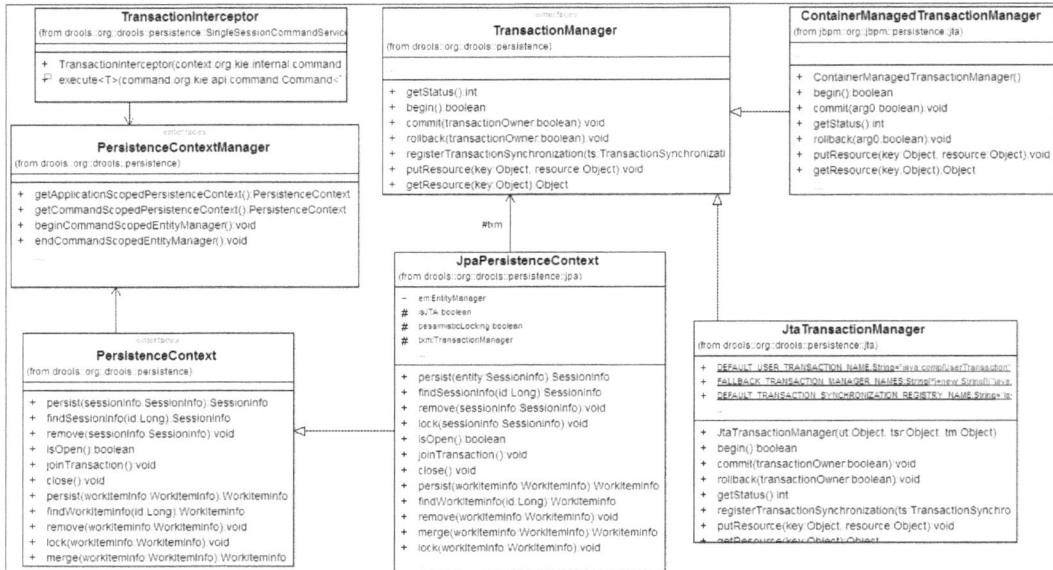

Summary

With this chapter, we took an extensive tour to the core engine components and services, not disregarding some of the engine inner implementation details. You should now be able to understand how the engine works and what happens "behind the curtains" when you use a specific engine feature.

The next chapter will deal with the engine customization and extension process in order to tailor the jBPM system features to your solution.

7
Customizing and Extending jBPM

This chapter details the extension points of jBPM. Not every user of jBPM uses the entire tool stack. Users will need to customize/extend jBPM to fit it into their solution architecture. This chapter will show you how jBPM's features can be customized and extended.

The list of topics that will be covered in the chapter is as follows:

- Domain-specific processes
- Writing your custom workitem handlers
- Customizing the process designer
- Extending variable persistence
- Extending user management

Custom process nodes

In *Chapter 5, BPMN Constructs*, we introduced the jBPM feature that lets you bind specific Java class implementations to the execution of a specific process task node type: send/receive tasks, service tasks, and ad hoc tasks.

These kinds of *extensible* task nodes are often called **custom workitems**, and the implementing classes that perform the horse work behind the process curtains are called **workitem handlers**. This architecture makes jBPM more flexible when it comes to adapting the engine to a particular domain, both in terms of features and tools UI. Let's start by reviewing the basics of the jBPM workitem and handlers.

Workitem and handlers

jBPM defines a work item as a unit of work that is defined inside the scope of a process but can be executed outside the engine; in particular:

- It accepts a set of parameters
- It performs some action
- It optionally returns a result

The workitem is just an abstract definition of a work unit and has several concrete, practical implementations in jBPM: human tasks, sendMessage tasks, and so on. The engine imposes no limitations to the workitem handler apart from enforcing the implementation of the `org.kie.api.runtime.process.WorkItemHandler` interface.

The engine runtime is instructed to bind a new handler implementation through the `WorkItemManager.registerWorkItemHandler(String workItemName, WorkItemHandler handler)` method, where the `workItemName` parameter must match a custom node name since it serves as the handler key.

jBPM itself extensively uses WorkItemHandler such as `LocalHTWorkItemHandler` (workitem name `Human Task`), `WebServiceWorkItemHandler` (workitem name `WebService`), or `RESTWorkItemHandler` (name `Rest`). This feature effectively streamlines the engine customization process, letting the user enhane (or replace) jBPM functionalities. You can find several jBPM workitem handler classes in the `jpbm-workitems-6.2.0.Final.jar` library (see package details in the following class diagram).

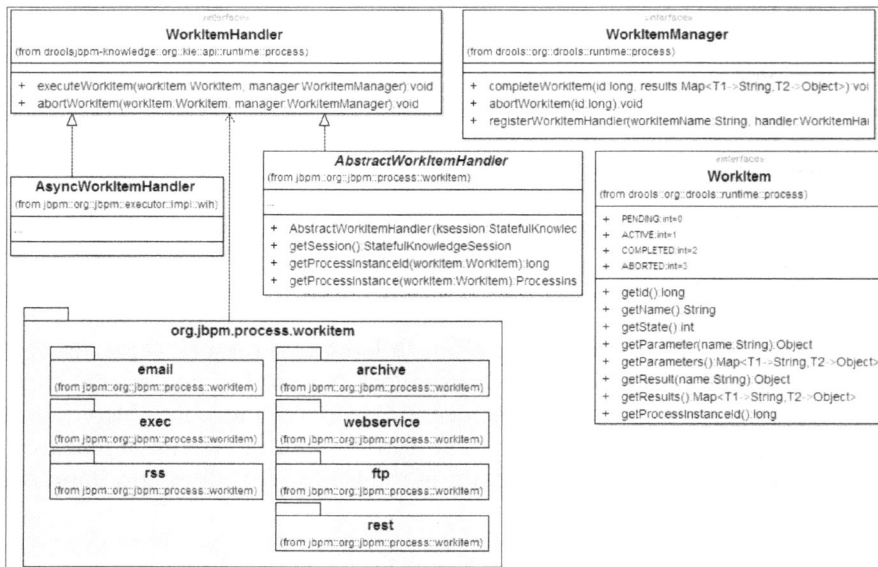

AsyncWorkItemHandler (we discussed it in the *Async task* section in *Chapter 5, BPMN Constructs*) can be found in the `jpbm-executor-6.2.0.Final.jar` library.

Life cycle

The workitem state transitions are as follows: ACTIVE, PENDING, COMPLETED, and ABORTED.

The WorkItemHandler call sequence is quite simple (see the following interaction diagrams), and when the handler calls complete or abort, the engine takes the control again and the process execution continues. The handler must implement two methods:

- executeWorkItem: The workitem manager invokes the executeMethod handler and, upon completion (at the end of executeMethod), the handler must invoke the callback method called completeWorkItem on the manager itself (optionally passing the output parameters):

- `abortWorkItem`: This method gets called as a consequence of a cancel or error event. The handler must perform clean-up operations (when needed) and call the manager back by the `abortWorkItem` method, which instructs the manager to set the workitem in the **ABORTED** state:

Cacheable and closeable interfaces

jBPM 6.2 introduces a new feature that lets the user hook into the workitem handler life cycle by implementing the following interface methods:

- `org.kie.internal.runtime.Closeable.close()`: Called on WorkItemManager (and Session) disposal. Here, you can perform the typical housekeeping duties (freeing resources, close connections, and so on).

- `org.kie.internal.runtime.Cacheable.close()`: Called when the jBPM internal cache manager is closed/disposed. By implementing the `Cacheable` interface, we enable our Workitem handler definition to be cached by jBPM.

> The RuntimeManager internally caches several configured object class definitions to optimize the initialization and startup times: event listeners, globals, marshalling strategies, Workitem handlers, and so on.

For a `Closeable` interface example, please have a look at the PizzaTweet handler implementation discussed in the following paragraphs.

Handlers in action

The workitem customization can be seen as a two-step process:

1. **Code implementation and runtime registration**: Make the handler implementation available to the runtime engine in order to trigger the handler execution when the engine reaches the custom node type

2. **Designer customization**: Enable the usage of the custom node from the UI interface (process designer)

Before diving into a detailed example (the `pizzatweet` project), let us look at the basics of the workitem architecture and review the three different ways in which we can register a handler: by direct registration with the API, by setting the `kmodule. xml` file, and by adding a handler configuration file.

Direct handler registration

The shortest, naïve way of registering a handler implementation with your engine session is to make a direct invocation to the Kie session's WorkItemManager:

```
// register MyWorkItemHandler for all ad hoc(None) task
ksession.getWorkItemManager().registerWorkItemHandler("task", new
MyWorkItemHandler());
```

This gives you a lot of flexibility; you do not need to define extra configuration files or properties (more on these in the upcoming section), and you are free to initialize your handler with everything you need during the execution. This is the preferred way to go when unit testing, particularly when replacing or defining a system workitem handler (`human task`, `service`, or `task`) since you do not have to tweak the Kie console to add the new node type to the **Service Task** menu, which would be mandatory to properly design the process definition.

Declarative kmodule.xml handler configuration

In case your project is a Kie module and you need a declarative, a less hardwired way to define the handlers is to add the `<workItemHandlers>` element to the `kmodule.xml` file as follows:

```
<kbase name="kbase" >
<ksession name="ksession">
<workItemHandlers>
  <workItemHandler name="pizzatweet"
  type="com.packt.masterjbpm6.pizzahandlers.PizzaTweetHandler">
</workItemHandlers>

</ksession>
```

Handler configuration file

When you need to add new custom node types, the preferred, standard way to register your handler implementation is to have it listed in the standard handler configuration file: the CustomWorkItemHandlers.conf file.

This file must contain the handler implementation class constructors and the work item name that will be used to register them; here you have, as an example, the default configuration file shipped with jBPM 6.2:

```
[
  "Log": new
  org.jbpm.process.instance.impl.demo.SystemOutWorkItemHandler(),
  "WebService": new
org.jbpm.process.workitem.webservice.WebServiceWorkItemHandler(kse
ssion),
  "Rest": new
org.jbpm.process.workitem.rest.RESTWorkItemHandler(),
  "Service Task" : new
org.jbpm.process.workitem.bpmn2.ServiceTaskHandler(ksession)
]
```

This file is written with the MVEL expression language and is loaded by the jBPM console runtime from the jbpm-console.war\WEB-INF\classes\META-INF folder; note that the filename is added to the sibling file called drools.session.conf whose content is as follows:

```
drools.workItemHandlers = CustomWorkItemHandlers.conf
```

> Note that from the system default handler definitions (Web Service and Service Task), by defining a constructor that accepts the ksession parameter, the KieSession will be automatically injected at runtime in your handler instance.

The very same property called drools.workItemHandlers is used to load the handler configuration file(s), for instance, during Kie Session initialization with KieSessionConfiguration. For example:

```
// create the session configuration
Properties props = new Properties();
props.setProperty("drools.workItemHandlers", "MyWorkItemHandlers.
conf");
KieSessionConfigurationconfig = KieServices.Factory.get().
newKieSessionConfiguration (props);
   // create the session
KieSessionksession = kbase.newKieSession(config,
EnvironmentFactory.newEnvironment());
```

Alternatively, with the runtime builder classes (see *Chapter 6, Core Architecture,* for details about runtime classes), you can have the following:

```
RuntimeEnvironmentBuilder.Factory.get().newDefaultBuilder()
    .addConfiguration("drools.workItemHandlers",
    "MyWorkItemHandlers.conf");
```

The `.conf` file is searched in the `META-INF/` classpath or in the `user.home` system folder.

> The property supports multiple space-separated entries such as the following:
> ```
> addConfiguration("drools.workItemHandlers",
> "MyWorkItemHandlers.conf OtherWorkItemHandlers.conf");
> ```

Handler definition file

The file that defines the WorkItemHandler process definition node properties is the workitem definition file (having the `.WID` extension), and it is written using the MVEL expression language.

When in the **Project Authoring** mode, the KIE console creates a default `WorkDefinitions.wid` file in your project `resources` folder that defines the `email`, `log`, `webservice`, and `rest` handlers whose custom nodes can be found under the Service Task menu of the process designer palette. Additional resources (custom node icons) are created in the `global` folder.

Here, you define your custom node properties and the workitem abstract definitions: `name` (used as the key for the handler), its parameters (input), result parameters (output), `displayName` (the label used for the node), and node icon resource. Let us have a look at our `PizzaTweet` custom node (we will talk about this in the example section):

```
[
    [
    "name" : "pizzatweet",
    "parameters" : [
  "tweetMsg": new StringDataType (),
  "tweetTags" : new ListDataType (),
  "tweetOrder" : new ObjectDataType
("com.packt.masterjbpm6.pizza.model.Order")
    ],
    "results" : [
```

```
        "details" : new ObjectDataType("java.util.Map"),
    "tweetOK": new BooleanDataType()
    ],
    "displayName" : "Pizza Tweet",
    "icon" : "../../../global/defaultservicenodeicon.png"
]
```

The `"parameters"` and `"results"` attributes are maps of typed parameters (name-type pairs).

- `parameters`: Defines the set of input parameters of the workitem
- `results`: Defines the set of output parameters

The types allowed for the parameters are the ones supported by the Drool core type classes: `StringDataType`, `IntegerDataType`, `FloatDataType`, `BooleanDataType`, `EnumDataType`, `ObjectDataType`, `ListDataType`, and `UndefinedDataType`. `ObjectDataType` wraps a type, while `ListDataType` wraps `java.util.List`.

Workitem handler implementation

Your handler must implement the `org.kie.api.runtime.process.WorkItemHandler` interface or extend the `AbstractWorkItemHandler` abstract class. This class gives you a set of useful helper methods in case your handler needs to get runtime information from the process or its nodes.

Handling input parameters

The handler can read the input parameters with the `WorkItem.getParameter (String name)` method. The input parameters are passed by the engine runtime, upon evaluation of your node data input set mappings.

Returning results to the process

The handler return object (`"results"`) is defined as a collection of parameters. Each attribute name must match a data output parameter in the `DataOutputSet` node settings (`"details"` and `"tweetOK"`), for example:

```
Map<String, Object> operationresults = new HashMap<String, Object>();
    operationresults.put("twitterCode", "200");
    results.put("details", operationresults);
    results.put("tweetOK", Boolean.TRUE);
```

The `details` and `tweetOk` keys must match both node data output parameter names (see the screenshot in the *Process and task parameter mappings* section) and the workitem handler `"results"` properties:

```
"results" : [
  "details" : new ObjectDataType("java.util.Map"),
  "tweetOK": new BooleanDataType()
]
```

The PizzaTweet example

The example guides you through the definition and the installation of a full working example project, which features a custom workitem that sends messages to Twitter (Twitter API integration not implemented for clarity).

> You can find a Twitter handler implementation in the Red Hat jBPM service repository: http://people.redhat.com/kverlaen/repository/Twitter/.

The main project (the `pizzatweet` KIE module) contains the process definition and the WID file, and it depends on custom types that are defined in the `pizzamodel` project (a plain utility JAR). These types are also used as parameter types by the `PizzaTweetHandler` handler (defined in the `pizzahandlers` project, another plain JAR).

The main project (the KIE module)

The main project depends on the two supplier projects: `pizzamodel` and `pizzahandlers`. Since these two projects are not KIE modules and do not need additional processing by the KIE runtime (they do not contain jBPM resources), we have set their Maven dependency scope to `provided` (see the `pom.xml` PizzaTweet project file). This speeds up the Maven build of our main KIE module; `kie-mave-plugin` in fact searches for KIE module dependencies only when their Maven scope is `runtime` or `compile`.

The KIE plugin for Maven (`kie-maven-plugin`) is the preferred way to build a KIE module. It ensures that all module business resources are valid at compile time and that the module can be successfully loaded at runtime. Make sure that you always have the `kie-maven-plugin` set in the `pom.xml` file of your KIE module (see the following PizzaTweet pom.xml excerpt):

```
<dependency>
  <groupId>com.packt.masterjbpm6</groupId>
  <artifactId>pizzamodel</artifactId>
```

```
      <version>1.0.0-SNAPSHOT</version>
      <scope>provided</scope>
    </dependency>
    <dependency>
      <groupId>com.packt.masterjbpm6</groupId>
      <artifactId>pizzahandlers</artifactId>
      <version>1.0.0-SNAPSHOT</version>
      <scope>provided</scope>
    </dependency>

    <build>
      <plugins>
        <plugin>
          <groupId>org.kie</groupId>
          <artifactId>kie-maven-plugin</artifactId>
          <version>6.2.0.Final</version>
          <extensions>true</extensions>
        </plugin>
      </plugins>
    </build>
```

Process and task parameter mappings

After having introduced the `pizzatweet` custom task definition in the previous section (the handler definition file), let us now look at how it fits into the process definition. The process definition is easy to understand; it has the custom tweet task and a script task that acts as a debugging step. The process variables (`msg`, `order`, and `tags`) are mapped to the custom task input parameters, while the resulting parameters (`tweetOK`, `details`) are mapped back to process variables (`success` and `results`). The following screenshot shows the **Assignments** properties panel for the `Pizza Tweet` custom node (see the `PizzaTweet` process definition):

	Assignment Type	From Object	Assignment Type	To Object
1	DataInput	msg	is mapped to	tweetMsg
2	DataInput	order	is mapped to	tweetOrder
3	DataInput	tags	is mapped to	tweetTags
4	DataOutput	tweetOK	is mapped to	success
5	DataOutput	details	is mapped to	results

After the tweet task completes, the script task is executed. As mentioned earlier, it simply dumps the updated process variables to the console for you to see. We are now going to look at the two dependent projects (handlers and models).

Handler project (pizzahandlers)

This project contains the handler implementation (the `PizzaTweetHandler` class) only, the one responsible for sending the tweet. In order to deploy the dependent `pizzahandlers` handler project, we have to perform a Maven "clean build install." The JAR file will then be installed in your system's Maven repository.

> Make sure that all the implementation classes and required dependencies are also available on the classpath of the application war (in this case, the war is our KIE console war), for example, by copying the required JAR files in the `/lib` folder.

Model project (pizzamodel)

The model project defines Java types for the process definition variables and the handler project parameters. In order to deploy the dependent `pizzamodel` project, we have to execute a Maven "clean build install." Thus, the JAR file is installed in your system's Maven repository so as to make it available to runtime dependency resolving.

IDE customization

In order to configure the KIE workbench (business process editor) tools and being able to use our custom node in the process editor, we have to create a workitem handler file. We create the `WEB-INF\classes\META-INF\PACKTworkItemHandlers.conf` file (in the `jbpm-console deployment` folder) and add the following content:

```
[
    "pizzatweet": new
com.packt.masterjbpm6.pizzahandlers.PizzaTweetHandler(ksession)
]
```

Then, we edit the `WEB-INF\classes\META-INF\drools.session.conf` file by adding our custom handler `.conf` filename to the `drools.workItemHandlers` property. Thus, the handler definitions from both the files are loaded. Please note that the handler configuration file names must be separated by a space:

```
drools.workItemHandlers = CustomWorkItemHandlers.conf
PACKTworkItemHandlers.conf
```

> The `drools.session.conf` file is picked up by the KIE console and read during KIE session initialization; see the *Handler configuration file* section for further details.

Copy the installed `pizzahandlers-1.0.0-SNAPSHOT.jar` and `pizzamodel-1.0.0-SNAPSHOT.jar` files to the jBPM console's `WEB-INF\lib` folder (for example, `wildfly-8.1.0.Final\standalone\deployments\jbpm-console.war\WEB-INF\lib`). This makes both the custom java types and the handler class available to the Kie console (a jBoss restart is required). Note that the Pizza Tweet (`name: pizzatweet`) custom task node is now displayed in the **Service Tasks** section of the object library:

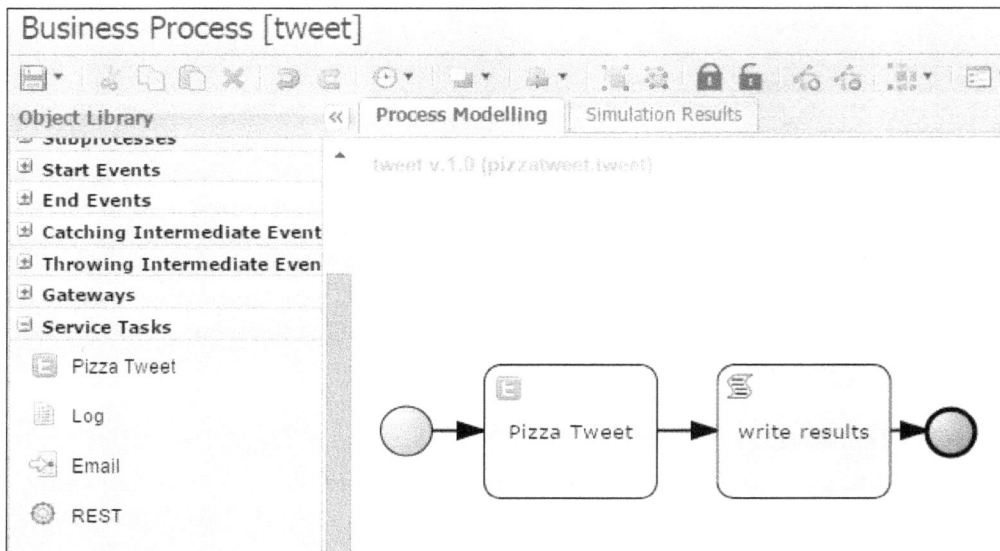

Console test run

As of jBPM 6.2.0 release, the KIE console is not much of a help in letting us test our process, since the generated task forms do not support complex type parameters automatically (our process takes an input parameter of the `Order` type); we cannot easily create our new process instances from here.

> The jBPM user guide (*Chapter 13, Forms*) explains the features shipped with the KIE console Form Modeler (`http://docs.jboss.org/jbpm/v6.2/userguide/chap-formmodeler.html`) and gives useful instructions on how to create customized human task forms and start process forms.

However, before leaving the console, let's check whether the process can successfully deploy without issues. Go to the `pizzatweet` project from the **Tools/Project Details** view and issue **build & deploy**. The process definition is registered with the runtime, and we should see it from the **Process Management/Process Definitions** tab.

Process Definitions			Details	
Name	Version	Actions	Definition Id	pizzatweet.tweet
Evaluation	1	⊕ Q	Definition Name	tweet
tweet	1.0	⊕ Q	Deployment	com.packt.masterjbpm6:pizzatweet:1.0
pizzadelivery	1	⊕ Q	Human Tasks	No Hum Tasks defined in this process
			Human Task Count	0
			User and Groups	No user or group used in this process
			Sub Processes	No subprocess declared by this process
			Process Variables	tags - java.util.List order - com.packt.masterjbpm6.pizza.model.Order success - Boolean msg - String
1-3 of 3			Services	Pizza Tweet - pizzatweet

Standalone test run

Get the PizzaTweetTest test class from the `PizzaTwitter` project and run (jUnit) the `newTweet` method:

```
// boilerplate code omitted for clarity;
// register the handler
session.getWorkItemManager().registerWorkItemHandler("pizzatweet",
newPizzaTweetHandler(session));
// init parameters

// start the process
ProcessInstanceinstance = session.startProcess("pizzatweet.tweet",
params);
```

The console prints the following text; first, we have the following handler log traces:

```
PizzaTweetHandler.executeWorkItem
PizzaTweetHandler.order=order: note=urgent cost=15.0
```

Then, we have the script task log traces, showing the following handler results:

```
tweet success:true
twitterCode:200
```

After the handler has been locally tested, we can move on and share it with the development team; this is where the service repository comes to the rescue.

Service repository

jBPM gives us the ability to add any handler to a public service repository; these are a collection of handler definitions that can be accessed both via HTTP or locally (the FILE protocol) so that handlers can be shared with the other developers.

At the time of writing this book, the KIE workbench supported two repositories: `http://people.redhat.com/kverlaen/repository` and `http://people.redhat.com/tsurdilo/repository`. Another repository service is available at `http://docs.jboss.org/jbpm/v6.2/repository/`. These repositories host several handler definitions; some of them are externally defined handlers (which means that the implementing JAR file is physically hosted in the repository), while other handlers are already defined in the jBPM runtime (for example, Java, REST, and transform handlers), and the repository just publishes the extended handler definition (`.WID`) file. The service repository is accessible through the **Connect to a service repository** button in the business process editor. Here, you have an example repository content dialog window:

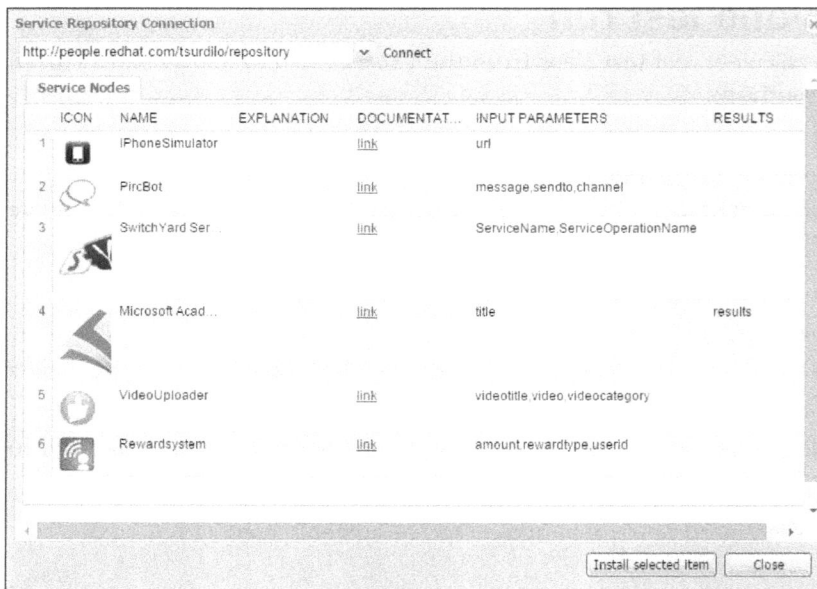

	ICON	NAME	EXPLANATION	DOCUMENTAT...	INPUT PARAMETERS	RESULTS
1		IPhoneSimulator		link	url	
2		PircBot		link	message,sendto,channel	
3		SwitchYard Ser..		link	ServiceName,ServiceOperationName	
4		Microsoft Acad...		link	title	results
5		VideoUploader		link	videotitle,video,videocategory	
6		Rewardsystem		link	amount,rewardtype,userid	

Service Repository Connection

http://people.redhat.com/tsurdilo/repository   Connect

Service Nodes

Install selected item   Close

We are going to see how to set up an additional custom local service repository.

> For additional details about the service repository, please refer to *Chapter 21, Domain-specific Processes* of the jBPM 6.2 user guide (*Service Repository* paragraph).

Custom service repository

A service repository is basically a folder structure containing handlers. The folder structure and the content to load are specified by a set of `index.conf` files. See the `repo.rar` example included with the book's source code.

Create a folder structure for our repository in a local folder (for example, `c:/temp/packtservicerepo`) containing the `pizzatweet` handler folder; inside the handler folder, we add the enhanced `pizzatweet.wid` file, which is basically a standard WID file with the following additional entries:

```
"defaultHandler" : "com.packt.masterjbpm6.pizzahandlers.
PizzaTweetHandler",
"documentation" : "index.html",
"category" : "examples",
"dependencies" : [
    "file:./lib/pizzahandlers-1.0.0-SNAPSHOT.jar",
    "file:./lib/pizzamodel-1.0.0-SNAPSHOT.jar"
  ]
```

The dependencies path is relative to the handler folder (`/lib`), and there we copy the two JAR files: the JAR file containing the handler definition and the JAR file defining the Java models.

> Refer to the details about the pizza handler and model projects in the `PizzaTweet` example paragraph.

It's worth mentioning that the WID file must have the handler folder name. After creating the files, we can open the service repository from the KIE workbench, giving the following local filesystem path: `file:///c:/temp/packtservicerepo`.

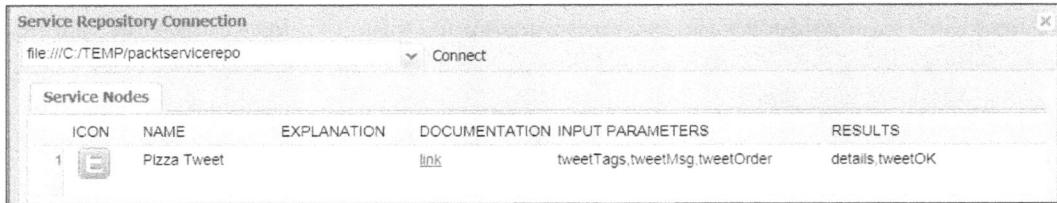

	ICON	NAME	EXPLANATION	DOCUMENTATION	INPUT PARAMETERS	RESULTS
1		Pizza Tweet		link	tweetTags,tweetMsg,tweetOrder	details,tweetOK

Service Repository Connection

file:///C:/TEMP/packtservicerepo ⌄ Connect

Service Nodes

Now, we can use **Install selected item** :. This makes the KIE workbench copy the assets to the internal KIE repository so that the handler becomes available.

jBPM identity management

In *Chapter 4, Operation Management,* we saw how the KIE workbench features JAAS-based user authentication and RBAC for the UI functionalities by means of the `user.properties` and `roles.properties` files.

The jBPM engine does not have built-in authentication or fine-grained authorization functionalities on process creation or task operations. TaskService and the human task management of users and groups with respect to task operations are delegated to a custom implementation of the `UserGroupCallback` interface. Here, the developer is able to implement his/her own task authorization mechanism by hooking into a custom identity management system or an ad hoc implementation.

jBPM provides a set of ready-to-use, configurable `UserGroupCallback` implementations:

- `DBUserGroupCallbackImpl`: Implementation that uses SQL queries to get user and group data from a database

- `LDAPUserGroupCallbackImpl`: LDAP system integration

- `MvelUserGroupCallbackImpl`: Default jBPM implementation when no callback is specified; the `UserGroupsAssignmentsOne.mvel` file is read and evaluated

- `JAASUserGroupCallbackImpl`: JAAS-based implementation to be used in a container (JBoss AS and Tomcat):

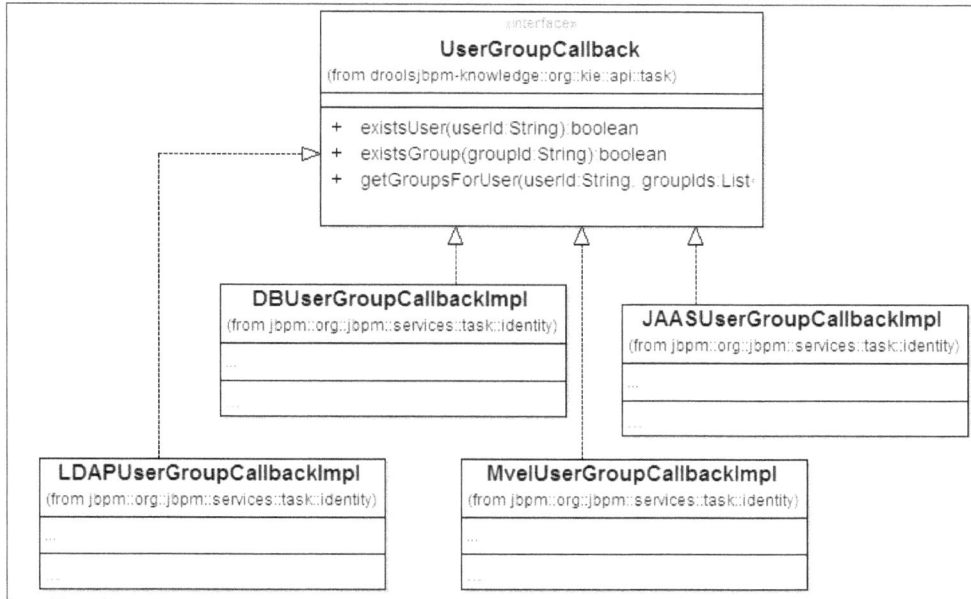

User callback configuration

The `UserGroupCallback` implementation is a singleton and can be set on the environment used to create `RuntimeEngine`:

```
// create the environment builder
RuntimeEnvironmentBuilder builder =
RuntimeEnvironmentBuilder.Factory.get().newDefaultBuilder()
.userGroupCallback(usergroupcallback);
// create the manager
RuntimeManagermanager = RuntimeManagerFactory.Factory.get()
.newSingletonRuntimeManager(builder.get(), "manager");
// create the runtimeEngine (omitted)
```

> All of the book's source code examples use a default custom callback class (`MyUserCallback`: you can find it in the `test-common` project). The callback is set by the `PacktJUnitBaseTestCase` class from which every test case borrows the runtime engine, the session, and so on.

The jBPM UserGroupCallback implementations rely on the `jbpm.usergroup.callback.properties` system property for the property filename or, alternatively, on the `jbpm.usergroup.callback.properties` property file for automatic configuration. The callback class can have a defined set of properties; let's review some of them for the classes shipped with jBPM.

The DBUserGroupCallbackImpl class

This callback has the following four properties (let `users` and `groups` be the tables defining our users and groups data):

- `db.ds.jndi.name`: JNDI name of the data source to be used, for example, `jdbc/jbpm-ds`.

- `db.user.query`: Query used to verify existence of the user (case-sensitive, expects a single parameter on position 1), for example:

  ```
  "select userId from users where userId = ?"
  ```

- `db.roles.query`: Query user to check group existence (case-sensitive, expects single parameter on position 1), for example:

  ```
  select groupId from Groups where groupId = ?"
  ```

- `db.user.roles.query`: Query used to get groups for a given user (case-sensitive, expects single parameter on position 1), for example:

  ```
  select groupId from Groups where userId = ?
  ```

The LDAPUserGroupCallbackImpl class

This callback relies on several properties (parameter descriptions):

- `ldap.bind.user` (optional if the LDAP server accepts anonymous access)
- `ldap.bind.pwd` (optional if the LDAP server accepts anonymous access)
- `ldap.user.ctx` (mandatory), for example, `ou\=Staff,dc\=packt,dc\=com`
- `ldap.role.ctx` (mandatory), for example, `ou\=Roles,dc\=packt,dc\=com`
- `ldap.user.roles.ctx` (optional; if not given, `ldap.role.ctx` will be used)
- `ldap.user.filter` (mandatory), for example, `=(uid\={0})`
- `ldap.role.filter` (mandatory), for example, `(cn\={0})`
- `ldap.user.roles.filter` (mandatory), for example, `(member\={0})`
- `ldap.user.attr.id` (optional; if not given, `uid` will be used)

- `ldap.roles.attr.id` (optional; if not given, `cn` will be used)

- `ldap.user.id.dn` (optional; is user id a `DN`?; instructs the callback to query for a user `DN` before searching for roles; default false)

- `ldap.search.scope` (optional; if not given, `OBJECT_SCOPE` will be used); the possible values are as follows: `OBJECT_SCOPE`, `ONELEVEL_SCOPE`, and `SUBTREE_SCOPE`

- `java.naming.factory.initial`

- `java.naming.security.authentication`

- `java.naming.security.protocol`

- `java.naming.provider.url`, for example, `ldap://localhost:10389`

jBPM serialization

We have seen that the engine features, with enabled persistence, state saving of session, process, task, and variable data to the database, and contextually, relevant object state data are marshalled and then, persisted on entity saving and unmarshaled on entity loading so as to make possible the preservation of the engine execution state in the long term, across system restarts. Here, the term **marshalling** is used because the jBPM serialization layer utilizes the Google Protobuf framework, a protocol, which, in the first instance, was used for RPC. Let us have a look at how the default jBPM serialization works and, later on, how we can manage to hook our serialization mechanism into jBPM.

Marshalling

`CommandService` (*Chapter 6, Core Architecture*) and the related interceptors are invoked to persist entities in a transactional way; the internal object marshalling phase takes place inside the transaction.

During the saving (or update) of the session, for example, its instance gets marshalled into its `RULEBYTESARRAY` column (`SessionInfo` table) so that the `Knowledge Session` instance can be loaded after a restart. The same happens for a process instance; its instance (with variables, node definitions, swimlanes, and so on.) is marshalled into `PROCESSINSTANCEBYTEARRAY` (the `ProcessInstanceInfo` table). The task data is marshalled into the `WORKITEMBYTEARRAY` (`WorkItemInfo` table).

The engine classes, which actually perform the marshalling job, are respectively `SessionMarshallingHelper` and `ProtobufProcessMarshaller`; these classes are internally baked by the `ProtobufMarshaller` class, which manages operations through a write handler (`ProtobufOutputMarshaller`) and an input handler (`ProtobufInputMarshaller`). The class diagram shown later demonstrates these classes and also how `KieMarshallers` (we introduced this in *Chapter 6, Core Architecture*) fits into the picture. It is simply a factory default for marshaller and strategy instances. `Strategies` are classes that control the marshalling process of variables.

Persisting variables

jBPM does not feature off-the-shelf processes and task variable persistence toward an ER model, mainly because of performance reasons. The main drawback is that you cannot search process instances by the value of a process variable. In order to add this kind of feature, you have to provide an implementation for the `JPAPlaceholderResolverStrategy` strategy (we are going to discuss it in a moment).

Strategies

As we have just seen, the default jBPM marshalling process results into bytes written in the database. This could be of limited interest to our application, but luckily, jBPM gives us the tools to hook into this mechanism by controlling the way variable (at least) marshaling works, by using or adding the so-called strategies.

During marshalling, in fact, jBPM delegates the serialization of process and task variables to strategy classes; jBPM ships with some ready-to-use strategies:

- `org.drools.core.marshalling.impl.SerializablePlaceholderResolverStrategy`: Features default Java serialization on objects implementing the `Serializable` interface. jBPM adds this strategy by default to the list of enabled strategies.

- `org.drools.persistence.jpa.marshaller.JPAPlaceholderResolverStrategy`: A strategy that manages variables as entities to and from a JPA-persistent store.

- `org.jbpm.document.marshalling.DocumentMarshallingStrategy`: This strategy manages marshalling for parameters of the `org.jbpm.document.Document` type. The document parameter type is used as the upload file parameter in the KIE Form modeler. These features are available with `jbpm-document-6-2-0.Final.jar`.

jBPM supports multiple strategies at once; it invokes them sequentially (a chain of responsibility patterns) following the order in which they are given when configuring the session (more on this in the following section). Each strategy (`ObjectMarshallingStrategy`) must specify the objects that it handles (the `accept` method) and provide the `marshal` and `unmarshal` methods (see the class diagram ahead).

Let us now look at how strategies can be configured using a working example which uses the jBPM `JPAPlaceholderResolverStrategy` in order to persist our process and task variables to our domain database table. Please refer to the `jbpm-marshalling` example project for a working marshalling example.

Configuring a marshalling strategy

The `jbpm-marshalling` example features a process definition (`rule_marshall. bpmn`), which uses an entity class as both the process variable and the task parameter. We want the engine to transparently persist our domain variable (the `OrderEntity` class) into a new domain database table (the `ORDERENTITY` table). The OrderEntity entity class must be added to our persistence unit (check the `persistence.xml` project), for example:

```
<class>com.packt.masterjbpm6.pizza.model.entity.OrderEntity</class
>
```

We set the marshalling strategies by passing an array of `ObjectMarshallingStrategy` to the environment that is used to create KieSession; in the following example (the `MarshallTest` example class), we configure the `JPAPlaceholderResolverStrategy` and the `SerializablePlaceholderResolverStrategy` strategies (please refer to the *RuntimeManager and the engine* section in *Chapter 6, Core Architecture*, for details).

```
RuntimeEnvironmentBuilder builder = RuntimeEnvironmentBuilder.Factory.
get().newDefaultBuilder();
builder.entityManagerFactory(super.getEmf())
builder.addEnvironmentEntry(
  EnvironmentName.OBJECT_MARSHALLING_STRATEGIES,
  new ObjectMarshallingStrategy[] {
    new CustomJPAPlaceholderResolverStrategy (super.getEmf()),
    new SerializablePlaceholderResolverStrategy(

    ClassObjectMarshallingStrategyAcceptor.DEFAULT) });
```

Note that we provided an extended `CustomJPAPlaceholderResolverStrategy`; this class extends and delegates all functionalities to the default `JPAPlaceholderResolverStrategy` and adds some logging features by sending the relevant information to the console during the marshalling process. Its constructor is given `EntityManagerFactory`, which has been created from the same persistence unit that the engine is using. This means that our entity table will be in the same engine database schema. The `ClassObjectMarshallingStrategyAcceptor` instance (used for `SerializablePlaceholderResolverStrategy`) performs the logic of accepting (filtering) object instances. Remember to always add the `SerializablePlaceholderResolverStrategy` strategy as the last strategy, since it's the one used by the engine.

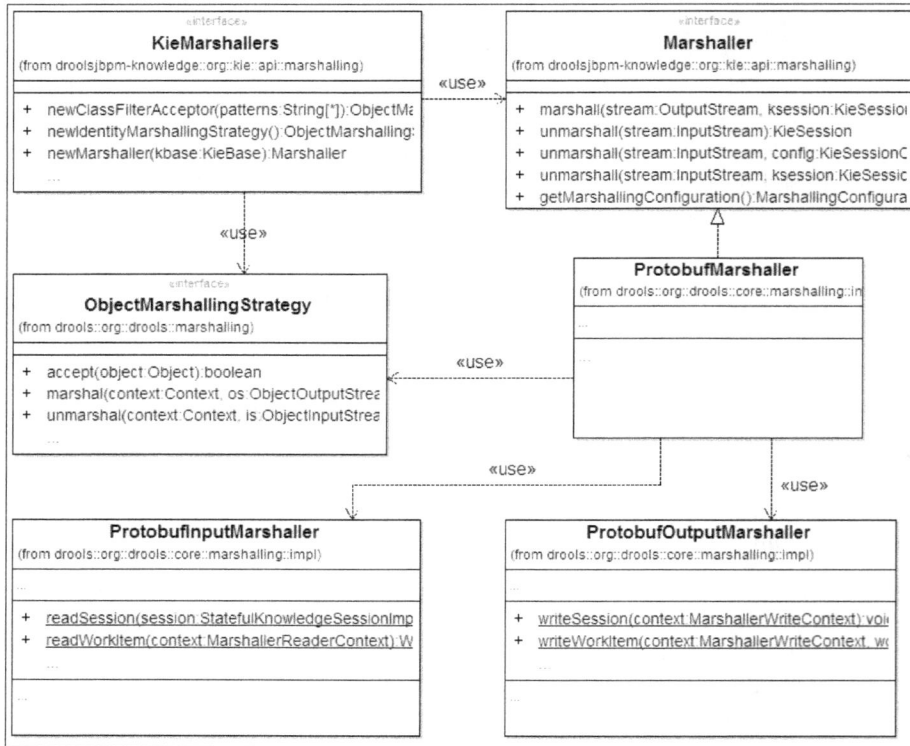

Persisting variables

The `MarshallTest` class is derived from `RuleTaskTest` (see the *Rule start event* section in *Chapter 5, BPMN Constructs*); it sets a global session variable and then, creates the process for passing two parameters, namely a plain `Order` instance and an `OrderEntity` instance, which are then passed to the user task. Upon the completion of the user task, the business rule is triggered and the global session variable is evaluated.

If we run the example jUnit test, we can see how the `marshall` method for our strategy class gets called several times before the human task is triggered:

```
18:19:42.757 [main] accepted
com.packt.masterjbpm6.pizza.model.entity.OrderEntity object:
[OrderEntity Id: null desc= First order amount=20.0]
18:19:42.757 [main] marshal
com.packt.masterjbpm6.pizza.model.entity.OrderEntity object:
[OrderEntity Id: null desc= First order amount=20.0]
18:19:42.788 [main] accepted
com.packt.masterjbpm6.pizza.model.entity.OrderEntity object:
[OrderEntity Id: 1 desc= First order amount=20.0]
18:19:42.788 [main] marshal
com.packt.masterjbpm6.pizza.model.entity.OrderEntity object:
[OrderEntity Id: 1 desc= First order amount=20.0]
18:19:44.318 [main] accepted
com.packt.masterjbpm6.pizza.model.entity.OrderEntity object:
[OrderEntity Id: 1 desc= First order amount=20.0]
18:19:44.318 [main] marshal
com.packt.masterjbpm6.pizza.model.entity.OrderEntity object:
[OrderEntity Id: 1 desc= First order amount=20.0]
18:19:44.350 [main] accepted
com.packt.masterjbpm6.pizza.model.entity.OrderEntity object:
[OrderEntity Id: 1 desc= First order amount=20.0]
18:19:44.350 [main] marshal
com.packt.masterjbpm6.pizza.model.entity.OrderEntity object:
[OrderEntity Id: 1 desc= First order amount=20.0]
```

The order entity is first inserted and then, updated several times; in the database table, we can see our record.

SELECT * FROM ORDERENTITY;			
ID	AMOUNT	DESCRIPTION	PIZZA
1	20.0	First order	null

It is the strategy's responsibility to maintain the entity state consistent between these calls. After the task completion, the unmarshall method gets called twice: first when the workitem is loaded from the database before its completion, and second when the process instance is loaded from the session:

```
luigi is executing task User Task 1
18:27:00.220 [main]
unmarshalcom.packt.masterjbpm6.pizza.model.entity.OrderEntity
object: [OrderEntity Id: 1 desc= First order amount=20.0]
18:27:00.251 [main]
unmarshalcom.packt.masterjbpm6.pizza.model.entity.OrderEntity
object: [OrderEntity Id: 1 desc= First order amount=20.0]
```

Summary

jBPM is open and configurable software. In this chapter, we reviewed three core features of the platform, which are commonly extended when tailoring jBPM systems to meet specific application requirements: domain processes and custom BPMN nodes, custom persistence for process and task variables, and human authorization based on custom implementation or legacy systems. The next chapter will provide the user with real-world jBPM solutions.

8
Integrating jBPM with Enterprise Architecture

We have an enterprise infrastructure in place, now we want to separate and centralize the process management to a single component, and of course, our choice is jBPM. So, the million dollar question would be "How do we integrate jBPM to the enterprise application in place?"

The answer to this question varies according to requirements and how the enterprise application is built. The architecture describes how the application is built, and from a broader perspective, a set of architecture patterns are used (either alone or in combination) as guidelines to model the architecture. This chapter focuses on provisions available in jBPM for integrating it with applications that follow these architecture patterns.

The chapter starts by discussing the context of enterprise application integration and continues to discuss the following in detail:

- Integrating jBPM into a JEE-based application
- Integrating jBPM into a service-oriented architecture
- Integrating jBPM into an event-driven architecture

Setting the context

System integration of a software component to an existing software architecture indicates that we should provide two windows (interfaces), listed as follows:

- To access services provided by the new component. In case of jBPM, it is represented by various services provided by jBPM, for example, the process runtime provision for managing the life cycle of a business process. JBPM exposes these services as APIs of its core engine.

- To enable jBPM to access the services provided by other components in the application architecture. The extension points that JBPM provides for integration with external components are the workitem handlers. We can create handlers and write the logic for accessing the external components.

The following figure depicts this context:

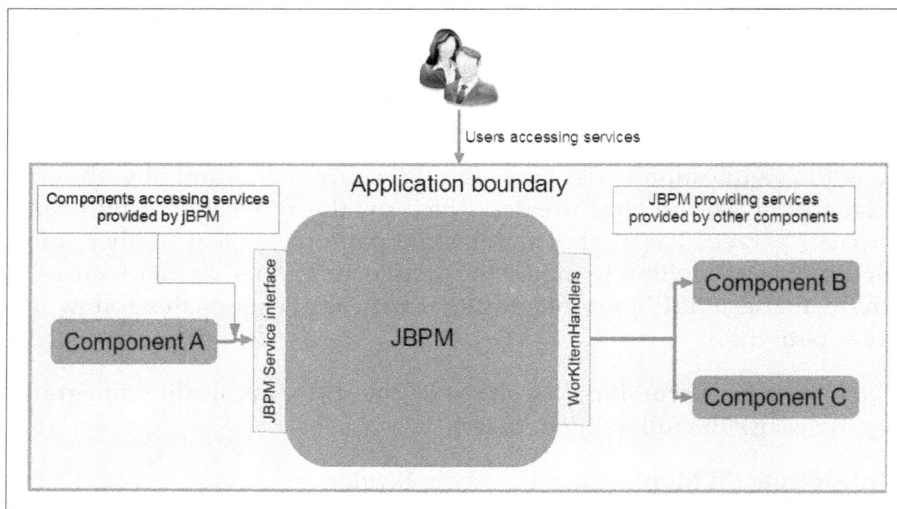

Services provided by jBPM

As we discussed in the previous section, one of the critical part of system integration with jBPM is the ability to access the features of jBPM. JBPM provides an application programming interface to access these features. This API can be directly invoked within the same JVM, and if needed to be accessed from outside the system boundary, it has to be wrapped and provided as a remotely accessible service. For this, we have an array of options, right from an **Enterprise JavaBeans** (**EJB**) remote interface to REST-based web services. Each of these will be detailed in the subsequent sections of this chapter.

The following are the services provided by jBPM:

- **Definition service**: This helps to define a process and analyze its content
- **Deployment service**: This helps to deploy a business process and the associated artifacts
- **Process service**: This helps to start a process instance from the process definitions, manage the life cycle of the instance, and interact with them using signals
- **User task service**: This helps to manage the human task life cycle
- **Runtime data service**: This helps to get the details of the data during jBPM runtime regarding process, process instance, tasks, and audit trails

Each service is detailed in the following section with (important) operations:

- `org.jbpm.services.api.DefinitionService`: This service helps to define a process from the BPMN text and provides operations to analyze a business process definition:

Operation	Operation signature	Description
buildProcess Definition	ProcessDefinition buildProcessDefinition(String deploymentId, String bpmn2Content, ClassLoader classLoader, boolean cache) throws IllegalArgumentException;	Builds the process definition from the given process definition content (bpmn2Content)
getReusable SubProcesses	Collection<String> getReusableSubProcesses(String deploymentId, String processId);	Gets the process identifiers of the reusable subprocesses inside a process definition
getProcess Variables	Map<String, String> getProcessVariables(String deploymentId, String processId);	Retrieves the name and type of all process variables in a business process
getServiceTasks	Map<String, String> getServiceTasks(String deploymentId, String processId);	Gets the identifiers of all service tasks associated in a business process definition
getTasks Definitions	Collection<UserTaskDefinition> getTasksDefinitions(String deploymentId, String processId);	Retrieves all the tasks defined in the business process

- `org.jbpm.services.api.DeploymentService`: This service helps to deploy and manage an application deployment unit:

Operation	Operation signature
`deploy`	`void deploy(DeploymentUnit unit);`
`undeploy`	`void undeploy(DeploymentUnit unit);`
`activate`	`void activate(String deploymentId)`
`deactivate`	`void deactivate(String deploymentId);`
`IsDeployed`	`boolean isDeployed(String deploymentUnitId)`

- `org.jbpm.services.api.ProcessService`: This process service is used to manage the life cycle and to interact with a started process instance:

Operation	Operation signature
`startProcess`	`Long startProcess(String deploymentId, String processId);`
`startProcess`	`Long startProcess(String deploymentId, String processId, Map<String, Object> params);`
`abortProcessInstance`	`void abortProcessInstance(Long processInstanceId);`
`abortProcessInstances`	`void abortProcessInstances(List<Long> processInstanceIds);`
`signalProcessInstance`	`void signalProcessInstance(Long processInstanceId, String signalName, Object event);`
`signalProcessInstances`	`void signalProcessInstances(List<Long> processInstanceIds, String signalName, Object event);`
`completeWorkItem`	`void completeWorkItem(Long id, Map<String, Object> results);`
`abortWorkItem`	`abortWorkItem(Long id);`

- `org.jbpm.services.api.UserTaskService`: This service helps to perform life cycle management operations of a user task:

Operation	Operation signature
`activate`	`void activate(Long taskId, String userId)`

Operation	Operation signature
claim	void claim(Long taskId, String userId)
Complete	void complete(Long taskId, String userId, Map<String, Object> params)
Delegate	void delegate(Long taskId, String userId, String targetUserId)
exit	void exit(Long taskId, String userId)
fail	void fail(Long taskId, String userId, Map<String, Object> faultData)
Forward	void forward(Long taskId, String userId, String targetEntityId)
release	void release(Long taskId, String userId);
resume	void resume(Long taskId, String userId);
skip	void skip(Long taskId, String userId);
start	void start(Long taskId, String userId);
stop	void stop(Long taskId, String userId);

- `org.jbpm.services.api.RuntimeDataService`: This API is used to retrieve information about the jBPM runtime including the data of process instances, tasks, and audit logs:

Operation	Operation signature
getProcesses	Collection<ProcessDefinition> getProcesses(QueryContext queryContext);
getProcessInstances	Collection<ProcessInstanceDesc> getProcessInstances(QueryContext queryContext);
getProcessInstance FullHistory	Collection<NodeInstanceDesc> getProcessInstanceFullHistory(long processInstanceId, QueryContext queryContext);
getVariableHistory	Collection<VariableDesc> getVariableHistory(long processInstanceId, String variableId, QueryContext queryContext);
getTaskEvents	List<TaskEvent> getTaskEvents(long taskId, QueryFilter filter);
getTasksOwned	List<TaskSummary> getTasksOwned(String userId, QueryFilter filter);

Creating custom workitem handlers

In order for jBPM to access services of other components in the application, we can use the workitem handler extension point provided by jBPM. Workitem handlers are used to specify domain-specific services to a BPMN activity. There are several inbuilt generic workitem handlers prebuilt in jBPM.

For creating a workitem handler, we have to implement the `org.kie.runtime.instance.WorkItemHandler` interface. This interface holds two methods to be implemented:

- `WorkItemManager.completeWorkItem(long workItemId, Map<String, Object> results)`
- `WorkItemManager.abortWorkItem(long workItemId)`

A custom workitem has to be registered to the engine by using the workitem manager. For example, for registering a customer task, we can use the following:

```
ksession.getWorkItemManager().registerWorkItemHandler("Notification",
new NotificationWorkItemHandler());
```

In conclusion, we have discussed the provisions available in jBPM for integrating it with generic software architecture. In the following sections, we will discuss how to integrate jBPM into widely used enterprise architectures.

Integrating with JEE

The Java Enterprise Edition provides an API and a runtime environment for developing and deploying enterprise applications. Further, EJB defines a set of lightweight APIs that can be used to build applications and leverage capabilities such as transactions, remote procedure calls, concurrency control, and access control.

EJB can be accessed in two modes:

- **Remote interface**: This is where the component that wants to access the EJB is not packed together with jBPM
- **Local interface**: This is where the component that wants to access the EJB is packed together with a jBPM service

JBPM provides out-of-the-box support for JEE integrations. It provides EJB remote and local interfaces for accessing the above-listed services.

EJB remote interfaces

The EJB remote interfaces are as follows:

Service name	EJB remote service class
Definition service	`org.jbpm.services.ejb.api.` `DefinitionServiceEJBRemote`
Deployment service	`org.jbpm.services.ejb.api.` `DeploymentServiceEJBRemote`
Process service	`org.jbpm.services.ejb.api.` `ProcessServiceEJBRemote`
Runtime data service	`org.jbpm.services.ejb.api.` `RuntimeDataServiceEJBRemote`
User task service	`org.jbpm.services.ejb.api.` `UserTaskServiceEJBRemote`

These remote services can be accessed from other Java applications. First, we need to access the `ejb` remote interface.

For example (specific to the `jboss` application server), the following code shows the lookup of the `ProcessService`:

```
final Hashtable<String, String> jndiProperties = new Hashtable<String,
String>();
  //Set the JNDI properties
  jndiProperties.put(Context.URL_PKG_PREFIXES,
  "org.jboss.ejb.client.naming");

  final Context context = new InitialContext(jndiProperties);
  //Set the bean name
  String beanName =
"ProcessServiceEJBImpl!org.jbpm.services.ejb.api.ProcessServiceEJB
Remote";

  String jndi = "ejb:/" + application + "/" + mappedName;
  ProcessService bean = (ProcessService) context.lookup(jndi);
```

After looking up the service, the service can be accessed seamlessly.

The EJB local interface

The EJB local interface can be accessed in two ways. One is by using the `javax.ejb.`EJB annotation and specifying the enterprise bean's local business interface name:

For example:

```
@EJB
ProcessService processservice;
```

The container will inject the EJB access for the API.

The other syntactical way to access a local EJB service is by using the JNDI lookup and the `javax.naming.InitialContext` interface's `lookup` method:

```
ProcessService processservice = (Processservice)
    InitialContext.lookup("java:jbpm/Processservice");
```

Integrating in SOA and EDA environments

The first part of this section covers how to integrate jBPM acting as a client into external services; *Chapter 5, BPMN Constructs*, and *Chapter 6, Core Architecture*, introduced jBPM elements specifically designed to call external web services from a process definition: the service task and the WS or the REST workitem handlers. The latter are jBPM ready-to-use, configurable components, but keep in mind that jBPM gives the user all the tools to develop custom handlers so as to perform interactions with generic external services (see *Chapter 7, Customizing and Extending jBPM*). The second part of the section will examine how to integrate the jBPM API as a server using REST, SOAP, and JMS. We will provide you with two example projects (`jbpm-remote-client` and `jbpm-remote-server`) in order to put into action these jBPM features.

We are going to see how to connect to both REST and SOAP services.

Integrating with REST services

Before starting a commented step-by-step tour of a jBPM application integrated with a REST service, let us review the basics of the support jBPM offers when coming to REST integration. The jBPM REST workitem handler (class `org.jbpm.process.workitem.rest.RESTWorkItemHandler`) is designed to interact with REST services (both secured and not secured); it supports the following parameters:

- `Url`: Target resource endpoint
- `Method`: HTTP method (defaults to `GET`)
- `ContentType`: Datatype when sending data (required with `POST` and `PUT`)
- `Content`: Data to send (required with `POST` and `PUT`)

- `ConnectTimeout`: Connection timeout (defaults to 60 seconds)
- `ReadTimeout`: Read timeout (defaults to 60 seconds)
- `Username`: Authentication username
- `Password`: Authentication password

The handler returns an output result that defines the following attributes:

- `Result`: The REST service text body response
- `Status`: The integer HTTP response code
- `StatusMsg`: A string description for the operation outcome

Our example application sets up a REST server and starts a process that has a REST service task node: the REST node performs an HTTP POST operation passing an `Order` instance (as a XML string) to the REST server; the server modifies the order's note and returns the order.

The REST service

Our REST server is started inside the test class (the `RestTest.initializeRestServer` method) by using the JAX-RS Apache CXF implementation (the CXF version is 2.7.14, check the project `pom.xml` file for dependencies); the initialization code sets a JAXB provider in order to support data binding for beans.

> Please check the Apache CXF documentation for JAX-RS at `http://cxf.apache.org/docs/jax-rs.html`.

The server is set up around a REST resource (the `RestResource` class), which defines the available operations through the JAX-RS `jax.ws.rs` package annotations.

The client – REST handler configuration

- The example test class method `RestTest.testRestProcess` starts a process instance (see the `rest.bpmn2` process definition); the process has a REST task node configured with the following mandatory parameters:

    ○ Url: `http://localhost:9998/pizzarestservice/order`

    ○ ContentType: `application/xml`

    ○ Content: `<order><note>my note</note></order>`

    ○ Method: POST

The node handler performs the REST call to the `postOrder(Order order)` method of the `RestResource` class; the method is annotated with `@Path("/order")`, and the XML bean serialization is taken care of, as we said, by JAXB. The REST task output variable is mapped back to the process instance and printed by the script task.

With the jUnit test class (`TestRest`), you can exercise the REST handler and the REST service outside the process definition (the `testPOSTOperation` method).

In case the default jBPM REST handler cannot meet your requirements (because of serialization constraints, frameworks lock-ins, and so on), it's important to point out that the developer can provide a brand new handler implementation: follow the *Chapter 7, Customizing and Extending jBPM*, guidelines describing the workitem handler development process. Let us now see how to set up and call a SOAP web service from a process definition.

The SOAP WebService

jBPM ships with a specialized `ServiceTaskHandler` (see *Chapter 5, BPMN Constructs*), which features web service interactions based on WSDL. The service task is marked as having an implementation of the `WebService` type (the task also supports plain Java implementation execution through the `Reflection` class). Please check the *Service task* section of *Chapter 5, BPMN Constructs*, for additional details and a working example description. Our jUnit class (`WsTest`) sets up a web service (the `startWebService` method) and then, starts a process that has two service task nodes, one calling the web service `addSmallOrder` operation, and the other calling the `addLargeOrder` operation: both the operations take an `Order` instance as the input and return a Boolean result, which is printed by the script task. The service tasks are on different process branches, which are taken by the exclusive gateway by evaluating the submitted order's total amount.

The JAX-WS service

The `TestWebService` service is an annotated JAX-WS service; it is started from the `WsTest.startWebService` method class, and its endpoint is set to `http://127.0.0.1:9931/testwebservice/order` (you can easily configure this in the unit test class). The `http://127.0.0.1:9931/testwebservice/order?WSDL` link returns the service WSDL interface. The service exposes the two aforementioned methods: `addSmallOrder` and `addLargeOrder`. Let us see how to call our web service operation from our process definition.

The client – process and service task handler configuration

In order to call the web service operation, we must perform the following steps, by editing the process definition and its service task node element:

Process definition

Things to be aware of are that we need to import the service WSDL definition. In the process definition import section, add the service WSDL location and namespace. The WSDL is passed to Apache CXF `JaxWsDynamicClientFactory`, which parses it when creating a dynamic web service client.

Editor for Imports				✕
Add Import				
Import Type	Class Name	WSDL Location	WSDL Namespace	
1 wsdl		http://127.0.0.1:9931/testwebservice/order?WSDL	http://ws.masterjbpm6.packt.com	⊘

Service task handler

The service task handler invokes the web service automatically by setting its parameters appropriately; this speeds up the integration process but may fall short when developing against service interfaces with complex types since, as we have already pointed out, the handler leverages the Apache CXF Dynamic Clients pattern. In this case, you are strongly suggested to develop a custom handler integrating your web service framework of choice. We set the handler parameters as follows:

- **Implementation**: `WSDL`
- **serviceInterface**: `TestWebService`
- **serviceOperation**: `addSmallOrder (addLargeOrder)`

The `mode` parameter value is left to `SYNC` (default), which translates in a blocking operation; when the `ASYNC` mode is set, the handler is forced to perform the web service call on a thread, returning the control to the process engine, and the workitem is completed as soon as the remote call returns.

The WebServiceWorkItemHandler class

jBPM offers a web service-oriented alternative to the service task handler with the `WebServiceWorkItemHandler` class. This handler improves over the service task handler in terms of parameter array handling, web service endpoint setting (it accepts the `Endpoint` parameter), and shortcut WSDL location loading (the `Url` and `Namespace` parameters instead of having to define the WSDL URL at the process definition level). The `serviceInterface` and `serviceOperation` parameters are renamed `Interface` and `Operation`, respectively.

jBPM as a remote service

The jBPM platform is offering a number of ready-to-use remote APIs in an effort to provide developers with an improved level of flexibility when designing solutions that require out-of-the-box jBPM integration. This remote service layer opens up a number of possibilities for providing the stakeholders with a flexible, open architecture, in order to satisfy and to quickly react to changing application requirements, for instance:

- A number of external application systems may require to occasionally connect to the jBPM runtime in order to check some task or retrieve some process information
- The jBPM operations manager may be constrained to perform administration tasks by submitting a batch of commands via HTTP only

jBPM ships with the following remote service interfaces:

- **REST API**.
- **JMS API**.
- **Java Remote API**: This API provides the developer with local stubs of the `KieSession`, `TaskService`, and `AuditService` core engine services. These service stubs of the API methods are wrappers for lower-level REST or JMS API calls.
- **SOAP API**.

All of these services are exposed by the jBPM KIE workbench, and as such, they are available only when the jbpm-console web application is deployed in a container.

> The source code for the remote services project is hosted at `https://github.com/droolsjbpm/droolsjbpm-integration/tree/master/kie-remote`.

The required Maven dependency for the remote service client is as follows:

```
<dependency>
  <groupId>org.kie.remote</groupId>
  <artifactId>kie-remote-client</artifactId>
  <version>6.2.0</version>
</dependency>
```

Let us now review the main remote service functionalities and how you can access them.

The REST API

This API provides functionalities in the following areas:

- **Runtime**: (`/runtime/` path) provides the user with process instance creation, process instance querying, and workitem operations

- **History**: (`/history/` path) provides with auditing data

- **Task**: (`/task/` path) provides task operation and task query methods

- **Deployments**: (`/deployments/` and `/deployment/` path) provides deployments management operations

For additional details, please check the jBPM user manual reference (*Chapter 17, jBPM Process Definition Language (JPDL)*).

Authentication

Upon invocation, the REST service operations check for the basic authentication user ID of your current HTTP session. For example, assume that you are performing REST operations in an unauthorized session by executing the following code from your command line:

```
curl -v http://localhost:8080/jbpm-console/rest/deployment/com.packt.
masterjbpm6:pizzadelivery:1.0
```

You will get an HTTP `401 Unauthorized` error (output edited for clarity; it may vary):

```
< HTTP/1.1 401 Unauthorized
< WWW-Authenticate: Basic realm="KIE Workbench Realm"
<html><head><title>Error</title></head><body>Unauthorized</body></
html>
```

> The Kie workbench default security mechanism leverages JAAS; the default configuration, for both jBoss WildFly and EAP, is stored in the application server XML configuration file (standalone and the like). See *Chapter 4, Operation Management*, for user and role configurations.

Otherwise, set the user ID and password (Workbench Realm) as follows:

```
http://admin:admin@localhost:8080/jbpm-console/rest/deployment/com.
packt.masterjbpm6:pizzadelivery:1.0
```

This will return the following response:

```
<deployment-unit>
  <groupId>com.packt.masterjbpm6</groupId>
  <artifactId>pizzadelivery</artifactId>
  <version>1.0</version>
  <kbaseName/>
  <ksessionName/>
  <strategy>SINGLETON</strategy>
  <status>DEPLOYED</status>
</deployment-unit>
```

> For a complete jBPM REST reference, please see the jBPM official documentation (*Chapter 17, Remote API*).

The remote Java API

The remote Java API is a high-level API that uses REST or JMS to interact with the remote engine services in order to provide the user with familiar service API classes (TaskService, KieSession, and so on).

Dependencies

The API depends on the jBoss RESTEasy REST implementation and the HornetQ JMS client library. The Maven dependency required to interact with the Remote API is, as we have pointed out earlier, the kie-remote-client module and the additional kie-remote-common artifact. Be sure not to have dependencies to the Apache CXF framework, which may cause issues with the jBoss RESTEasy framework.

The REST client

The initialization is done with a builder **fluent** API obtained from RemoteRuntimeEngineFactory:

```
// the deploymentId identifies the KIE module
public static String deploymentId =
"com.packt.masterjbpm6:pizzadelivery:1.0";
RemoteRestRuntimeEngineBuilder restEngineBuilder =
RemoteRuntimeEngineFactory.newRestBuilder()
.addDeploymentId(deploymentId)
.addUrl(instanceurl).addUserName(user)
.addPassword(password);
RemoteRestRuntimeEngineFactory engineFactory = restEngineBuilder
```

```
.buildFactory();
// get the engine
RemoteRuntimeEngine engine = engineFactory.newRuntimeEngine();
// and the services
TaskService taskService = engine.getTaskService();
KieSession ksession = engine.getKieSession();
ProcessInstance processInstance = ksession.startProcess(processID);
```

> Please see the `jbpm-remote-server` Maven project and its `RestTest` jUnit class for a full working example.

Client for jBPM JMS service

When using the JMS Remote API client, we need to add a number of library dependencies, notably HornetQ and the jBoss remote client. We are going to see how to configure and run a remote client application, which creates a jBPM process instance.

> Please see the `jbpm-remote-server` Maven project and its `JmsTest` jUnit class for a full working example (WildFly 8.1 is required to be up-and-running).

Server JMS configuration

WildFly comes with HornetQ as JMS MQ middleware; in order to have JMS to properly work, we need to check the jBPM JMS queues are registered with the JNDI service and that the user security settings are set. By default, HornetQ will use the "other" JAAS security domain, which is the one used by KIE Workbench Realm for authentication (recall the `user.properties` and `roles.properties` files). In addition, HornetQ defines authorization settings in the following element of the WildFly `standalone-full.xml` configuration file (under the messaging subsystem):

```
<security-settings>
  <security-setting match="#">
    <permission type="send" roles="admin guest"/>
    <permission type="consume" roles="admin guest"/>
    <permission type="createNonDurableQueue" roles="admin
    guest"/>
    <permission type="deleteNonDurableQueue" roles="admin
    guest"/>
  </security-setting>
</security-settings>
```

Here, we just add the KIE console `admin` role (along the default `guest` role); the `admin` role is already configured for JAAS.

Now, to check whether our JMS user is properly configured, open the jBoss management console (`http://localhost:9990/console`) and select **Configuration/Subsystems/Messaging/Destinations** and select **Default Provider** and **Security Settings** on the top navigation bar; you shall view the defined users.

The WildFly jBPM JMS queue configuration is defined in the `jbpm-console.war\WEB-INF\bpms-jms.xml` file; the remotely accessible queues are registered in the `java:jboss/exported` JNDI namespace.

To check whether the jBPM JMS queues are correctly bound to JNDI, open the jBoss management console (`http://localhost:9990/console`) and select **Runtime/Status/Subsystems/JNDI View**; here, you shall view the **KIE.AUDIT**, **KIE.SESSION**, **KIE.RESPONSE**, and **KIE.TASK** queues. Here, you should also have `RemoteConnectionFactory` listed; this factory allows for remote connection to the jBoss JNDI namespaces (we are going to see this in a moment).

```
▼  java:jboss:exported

  ▼  jms

      RemoteConnectionFactory                                    HornetQConnectionFactory
                                                                 [serverLocator=ServerLoca ...

    ▼  queue

        KIE.AUDIT                                                                   AUDIT]

        KIE.RESPONSE                                                             RESPONSE]

        KIE.SESSION                                                               SESSION]

        KIE.TASK                                                                     TASK]
```

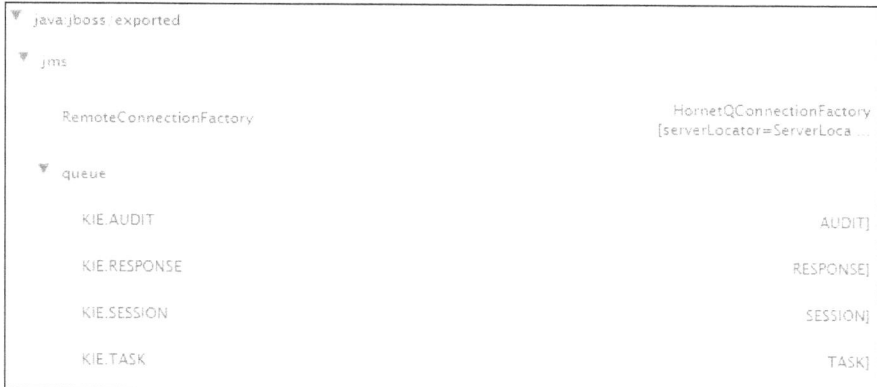

> For the WildFly messaging security configuration, please refer
> to https://docs.jboss.org/author/display/WFLY8/
> Messaging+configuration. For the official HornetQ reference,
> please see the latest documentation at http://docs.jboss.
> org/hornetq/2.4.0.Final/docs/user-manual/html.

JMS client implementation

To set up a remote JMS client connection, we use the same approach that we used
for the REST client; we configure a specialized builder, provided by the good old
RemoteRuntimeEngineFactory.

> Please see the jbpm-remote-server Maven project and its
> JmsTest jUnit class for the full working example.

```java
// the deploymentId identifies the KIE module
public static String deploymentId = "com.packt.
masterjbpm6:pizzadelivery:1.0";
/* the remoteInitialContext is an instance of the jBoss Naming
 service (InitialContext) and gives you access to the container
 remoting services for JMS */
/* the connectionfactory represents the JMS connection configuration
settings */

RemoteJmsRuntimeEngineBuilder jmsEngineBuilder =
RemoteRuntimeEngineFactory
  .newJmsBuilder().addDeploymentId(deploymentId)
  .addRemoteInitialContext(remoteInitialContext)
  .addUserName(jms_user).addPassword(jms_password)
  .addConnectionFactory(connectionfactory)
  .addTimeout(maxTimeoutSecs);
```

We get the factory from the builder and the engine from the factory:

```
RemoteJmsRuntimeEngineFactory engineFactory = jmsEngineBuilder
  .buildFactory();
RuntimeEngine engine = remoteJmsFactory.newRuntimeEngine();
```

Then, we get the service classes from the engine:

```
TaskService taskService = engine.getTaskService();
```

In order to make the jBPM remote client resolve the remote jBPM queues, we need to configure the jBoss JNDI provider URL as follows:

```
initialProps.setProperty(InitialContext.PROVIDER_URL,

"http-remoting://" + jbossServerHostName + ":8080");
```

WildFly utilizes an HTTP upgrade and features port multiplexing for nearly all of its protocols. The jBoss remote JNDI historically listened on port 4447, but is now, on port 8080.

> For a complete WildFly reference, please see `https://docs.jboss.org/author/display/WFLY8/Documentation`.

The SOAP API

The jBPM workbench features additional interoperability by exposing a SOAP service as described by the `/jbpm-console/CommandService?WSDL` endpoint; the service implements a single `execute` operation. At the time of writing this book, the WSDL available for the jBPM 6.2.0 release could not be used to generate the client classes because of some typos in WSDL.

The client Maven dependency is as follows:

```
<dependency>
  <groupId>org.kie.remote.ws</groupId>
  <artifactId>kie-remote-ws-common</artifactId>
  <version>6.2.0.Final</version>
</dependency>
```

For the sake of completeness, we will now describe how to call into jBPM by using its SOAP API. Our `jbpm-remote-server` test project, the `SOAPTest` jUnit test class, creates a web service client and then, starts a new process instance.

First, we get the WSDL resource from the endpoint URL as follows:

```
URL commandWsdlUrl = new URL(
"http://localhost:8080/jbpm-console/CommandService?WSDL");
```

The `execute` command operation accepts the `JaxbCommandsRequest` command, which is a DTO (serializable) wrapper for plain jBPM command classes (see *Chapter 6, Core Architecture*). All jBPM command classes are also JAXB-annotated classes.

```
StartProcessCommand startProcessCommand = new
StartProcessCommand();
startProcessCommand.setProcessId(processID);
JaxbCommandsRequest request = new
JaxbCommandsRequest(deploymentId, startProcessCommand);
```

`JaxbCommandsRequest` can also accept a batch of commands unlike REST or the JMS remote API.

Transactions

When calling into jBPM by using REST, SOAP, or the remote Java API, you are in control of your transaction management. If the jBPM call is supposed to be part of a transaction and this call fails or throws an exception, you must handle it and perform rollback operations or compensate the business logic on your side.

All the remote API methods throw a `RemoteApiException` exception to indicate that the remote call (either REST or JMS) has failed.

The SOAP API `execute` operations throws `CommandWebServiceException`. If you need a tight integration and transaction propagation mechanism, you should consider moving to an EJB layer wrapping the full-fledged jBPM services (see the *Integrating with JEE* section at the start of this chapter).

Summary

In this chapter, we expanded on the jBPM features targeted at enterprise architecture integration. We discussed the core services exposed by jBPM and how they can be accessed by using different technologies such as JEE, SOAP, REST, and JMS.

In the next chapter, we will focus on details that have to be taken care of while deploying jBPM to production.

9
jBPM in Production

In the previous chapters, we sailed through the various functional aspects of jBPM and also saw how we can extend and customize jBPM for adding more features. Now, it is production time, and there is a change in the perspective of how the application is viewed by its stakeholders.

The important question now is not the functional characteristics, but the nonfunctional ones. People think about the stability and resilience of the application and not the flexibility that it gives. People think of how fast and cost-efficiently the application can be scaled so as to provision for more users and how less critical is the latency of a service.

The mettle of the application is put under fire. jBPM is ready-to-use production software, and in this chapter, we will discuss various facilities available in jBPM to make it fit into the requirements of production software. The chapter is structured on the basis of the major qualities of a system that has to be taken care of in production.

We will discuss the following topics:

- How to scale
- How to make applications secure
- How to meet availability requirements
- How to incorporate new changes to the system
- How the system handles errors in runtime

Scalability

Scalability can be described as the capacity of a system to handle growing volumes of service provisioning in a controlled and cost-efficient manner. In case of a BPM system, there are two major use cases where the requirement of scaling can arise.

- Scaling the modeling facility, that is, the workbench
- Scaling the process runtime, with which the end customers of the application interact

Scaling an application typically involves two methods:

- **Vertical scaling**: This is achieved by adding resources to the server that is providing the service
- **Horizontal scaling**: This is achieved by adding multiple servers to provision the same service

Vertical scaling involves less complexity of implementation as it asks for improving the hardware (usually) and configuring the application to use these resources. However, vertical scaling is often limited by the constraints put by cost and technology in building the resources. In the context of jBPM, the resources that can be added are the memory, processor cores, and secondary storage mechanisms. jBPM doesn't provide out-of-the-box functionality to explicitly cater for these resource improvements, but there would be improvements in throughput and performance by taking advantage of the underlying platforms used by the jBPM, such as the application server on which jBPM is deployed and JVM on which the application server resides.

Within the scope of this book, it is obvious that horizontal scaling would need a better method, and the following sections purely concentrate on horizontally scaling of the jBPM functionality.

Scaling the business process modeling facility

Scaling the modeling tool points to increasing the number of users that can perform modeling simultaneously. Users can choose either web tooling or Eclipse tooling for modeling purposes, and there can be scenarios where the modeling users create a single application or multiple applications.

Given the previous factors and constraints, the most obvious way to increase the throughput of the modeling workbench service is to increase the number of units that provision the service. So, we add servers and jump to face the quintessential problems of clustering. Each server has a separate asset repository, and if the users collaborate to create the same application, we need to keep the assets in the repository sync always.

The out-of-the-box facility that jBPM provides as the asset repository is the Git-backed **Virtual File System (VFS)**, and in order to keep the file system in sync, jBPM suggests the use of Apache Helix, a cluster management framework.

The following diagram visualizes the deployment architecture in this scenario:

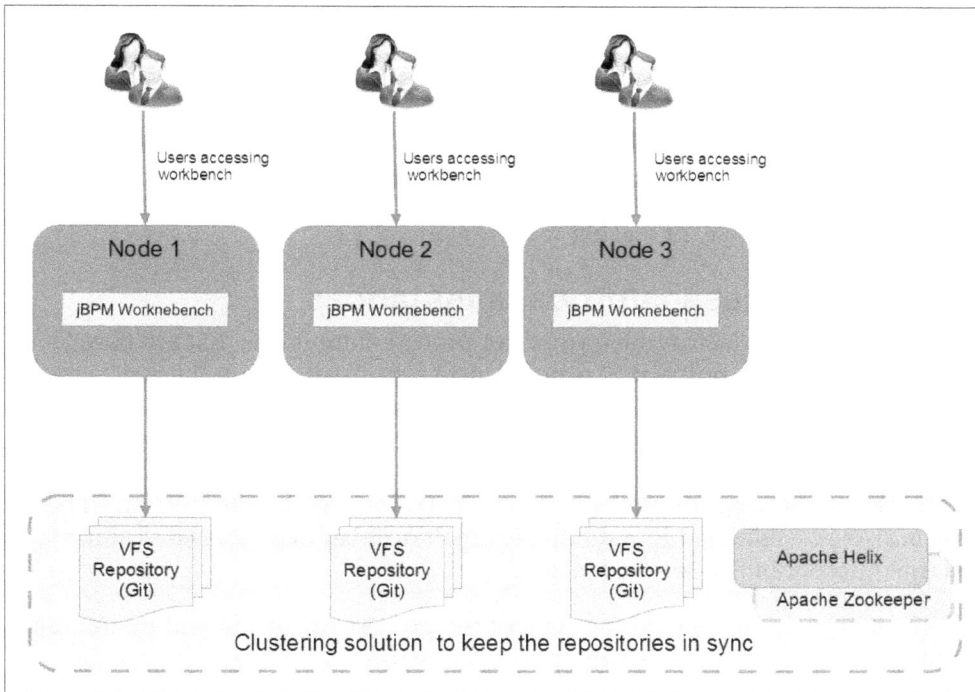

Apache Helix acts as a cluster management solution, which registers all servers to the cluster and enables the synchronization of the repositories.

Helix internally uses Apache ZooKeeper to manage the state of the system and manage the notifications between the nodes.

The details of configuring VFS clustering are explicitly provided in the jBPM user guide; please refer to it for configuration details.

Apache Helix provides a set of functionalities that enable us to develop a fault-tolerant, scalable distributed system. For more details, see `http://helix.apache.org/Architecture.html`.

Scaling the process runtime

When we talk about scaling any software application, it involves increasing the capacity of the system to serve an increasing number of user interactions. In BPM-based applications, along with the increase in user interactions, the complexity and content of the business processes factor in for an increase in the capacity of the system.

Factors and considerations

The following section highlights the factors involved in finalizing the deployment architecture of the system and discusses the consideration made in the jBPM architecture to meet these increasing requirements.

Number of processes/process instances

Yes, this is an obvious factor: the number of process definitions that are part of the application and the number of process instances created from these process definitions use the system capacity. The process definitions are stored in the asset repository, and we have already discussed this in the *Scaling the business process modeling facility* section, but the increase in process definitions directs to an increase in the number of process instances that have to be managed by the system. The reverse is also possible, that is, a relatively small number of process definitions but a large number of process instances.

Process instances carry the runtime state of the process execution and by default in memory. However, this is not an option in a real-world scenario where the availability of the runtime state is critical, and so, jBPM provides mechanisms to persist the process instances into a database. In the context of our discussion, we have to note that with an increase in the number of process instances, we have to do the following:

- Increase the capacity of the database
- Increase the capacity of the memory

The following diagram shows the schematic deployment architecture where there are multiple jBPM runtime instances having replicated VFS repositories for asset storage and a centralized database storing the runtime information:

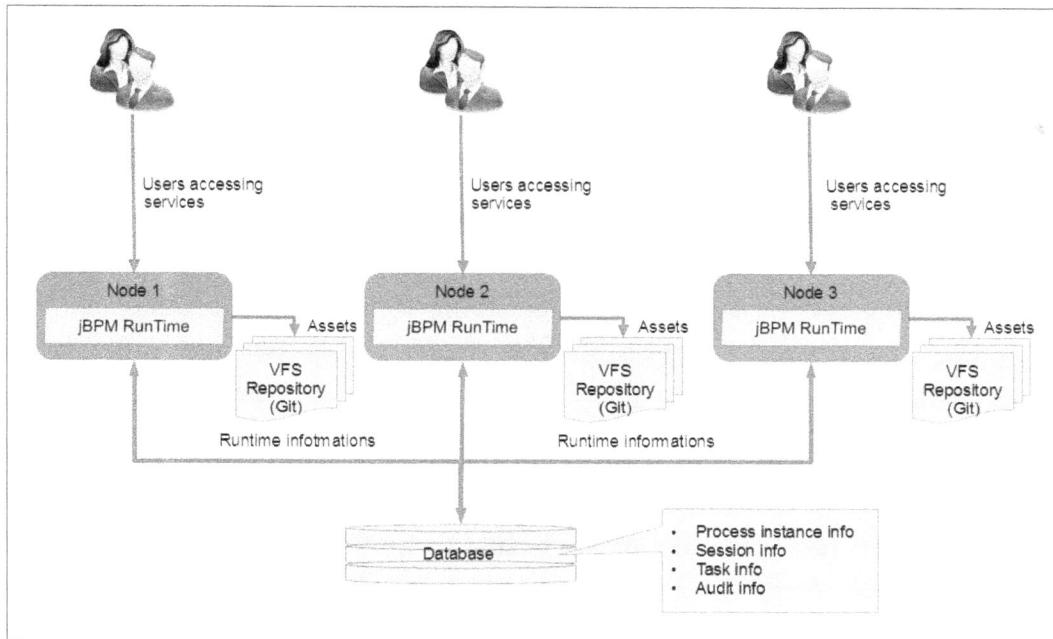

The number of user interactions

User interactions with the process runtime are for the following:

- Interactions with the process engine for starting a process (creation of a process instance) or signaling an event to an already started process instance

- Interaction with the task service for handling the life cycle methods for human tasks

> One other possibility is the interactions to domain-specific asynchronous workitems, which needs its own capacity planning.

Interactions with the process engine are performed through a dedicated KieSession API and come with a specific constraint; that is, interactions with a process instance can only be performed through the Kie session in which the process instance was created. Each interaction needs the instance of the Kie session where it was created, and jBPM provides multiple strategies for handling the scaling up of process interactions. jBPM allows you to choose the strategy while creating the runtime manager, which is in turn used to access the runtime engine and create sessions for interaction.

Flavors of runtime manager

- **Singleton strategy**:

  By choosing this strategy, we choose to maintain a single instance of the runtime instance and a single instance of Kie Session for all interactions. This is the easiest strategy and is most favorable for low and medium loads.

  The singleton strategy can be programmatically chosen as follows:

  ```
  RuntimeManagerFactory.Factory.get().newSingletonRuntimeManager(run
  timeEnvironment);
  ```

  Further, if you are using jBPM console (KIE Workbench), you can configure the `<runtime-strategy>` deployment descriptor tag as SINGLETON. A sample deployment descriptor is shown as follows:

  ```
  <deployment-descriptor xsi:schemaLocation="http://www.jboss.org/
  jbpm deployment-descriptor.xsd" xmlns:xsi="http://www.w3.org/2001/
  XMLSchema-instance">
    <persistence-unit>org.jbpm.domain</persistence-unit>
    <audit-persistence-unit>org.jbpm.domain</audit-persistence-
    unit>
    <audit-mode>JPA</audit-mode>
    <persistence-mode>JPA</persistence-mode>
    <runtime-strategy>SINGLETON</runtime-strategy>
    <marshalling-strategies/>
    <event-listeners/>
    <task-event-listeners/>
    <globals/>
    <work-item-handlers/>
    <environment-entries/>
    <configurations/>
    <required-roles/>
  </deployment-descriptor>
  ```

  SINGLETON is the default strategy in the jBPM console.

We can override the default deployment descriptor by using a Java option during the start of the server as follows: `Dorg.kie.deployment.desc.location=file:/application/configuration/deployment-descriptor.xml`.

- **Per request strategy**:

  A new instance of runtime manager is provided, and the session is created and maintained for the request scope.

  This strategy is stateless and ideal for horizontally scaling the jBPM process runtime instances, but the functionality in the process is limited to stateless facts, with no user interactions allowed other than the start process.

  We can choose the per request strategy programmatically during the creation of runtime by the following code:

  ```
  RuntimeManagerFactory.Factory.get().newPerRequestRuntimeManager(runtimeEnvironment);
  ```

  Further, for jBPM console, the deployment descriptor can be customized as follows:

  ```
  <?xml version="1.0" encoding="UTF-8" standalone="yes"?>
  <deployment-descriptor xsi:schemaLocation="http://www.jboss.org/jbpm deployment-descriptor.xsd" xmlns:xsi="http://www.w3.org/2001/XMLSchema-instance">
    <runtime-strategy>PER_REQUEST</runtime-strategy>
  </deployment-descriptor>
  ```

- **Per process instance strategy**:

  It is the most advanced strategy taking into consideration the tradeoff between the scalability of the system and the overhead it put. As the name indicates, ksession sticks to the process instance and lives as long as the process instance is alive. It does not have the scalability constraints in the singleton strategy, and although it has a high overhead, it doesn't have the overhead limitations and is not scalable as the per request strategy. Thus, the per process instance strategy is placed in the middle of the above two and is used in most of the cases where jBPM is used.

  We can choose the per process instance strategy programmatically during the creation of runtime by the following line of code:

  ```
  RuntimeManagerFactory.Factory.get().newPerProcessInstanceRuntimeManager(runtimeEnvironment);
  ```

Further, for the jBPM console, the deployment descriptor is customized as follows:

```
<?xml version="1.0" encoding="UTF-8" standalone="yes"?>
<deployment-descriptor xsi:schemaLocation="http://www.jboss.org/
jbpm deployment-descriptor.xsd" xmlns:xsi="http://www.w3.org/2001/
XMLSchema-instance">
  <runtime-strategy>PER_PROCESS_INSTANCE</runtime-strategy>
</deployment-descriptor>
```

Task service

Task service is the dedicated component for managing human task services. The interactions with an application can be performed through the human tasks. jBPM provides a default implementation for the human task service, which is based on the WS-Human Task specification. The task service clients are light and go hand-in-hand with strategies that we have chosen for the runtime manager, and all the clients share the same database; therefore, scaling up of human tasks is in sync with the strategy chosen for the runtime manager and with the increase in the capacity of the database storage.

Number of timer events

The functionality of timer events is achieved using the scheduler service. Multiple implementations of the scheduler service are provided by jBPM. The Quartz scheduler-based implementation is a fit in for a production environment. The Quartz scheduler supports the clustering mode, which provides both high availability and scalability, which works by maintaining the data or state of the schedules (or jobs) that it handles in a shared database between nodes.

> Quartz is an open source job scheduling library that can be integrated within a Java application. Quartz can be used to create scheduled tasks and provide support for JTA transactions and clustering. For more details, see http://quartz-scheduler.org/.

The Quartz scheduler can be enabled by providing the absolute path of the quartz definition file against the org.quartz.properties system property.

A sample quartz definition file is given as follows that is configured for use along with a PostgreSQL database.

```
#========================================================================
========
# Configure Main Scheduler Properties
```

```
#=====================================================================
=======

org.quartz.scheduler.instanceName = jBPMClusteredScheduler
org.quartz.scheduler.instanceId = AUTO


#=====================================================================
=======
# Configure ThreadPool
#=====================================================================
=======

org.quartz.threadPool.class = org.quartz.simpl.SimpleThreadPool
org.quartz.threadPool.threadCount = 5
org.quartz.threadPool.threadPriority = 5


#=====================================================================
=======
# Configure JobStore
#=====================================================================
=======

org.quartz.jobStore.misfireThreshold = 60000

org.quartz.jobStore.class=org.quartz.impl.jdbcjobstore.JobStoreCMT
org.quartz.jobStore.driverDelegateClass=org.quartz.impl.jdbcjobstore.
PostgreSQLDelegate
org.quartz.jobStore.useProperties=false
org.quartz.jobStore.dataSource=managedDS
org.quartz.jobStore.nonManagedTXDataSource=notManagedDS
org.quartz.jobStore.tablePrefix=QRTZ_
org.quartz.jobStore.isClustered=true
org.quartz.jobStore.clusterCheckinInterval = 20000
#=====================================================================
=======
# Configure Datasources
#=====================================================================
=======
org.quartz.dataSource.managedDS.jndiURL=jboss/datasources/psjbpmDS

org.quartz.dataSource.notManagedDS.jndiURL=jboss/datasources/
quartzNotManagedDS
```

> When using the Quartz scheduler, as a prerequisite, we would have to create the database schemas that are used by Quartz to persist its job data. The database scripts provided with the Quartz distribution (jBPM uses Quartz 1.8.5. DB scripts) are usually located under `QUARTZ_HOME/docs/dbTables`.

The scheduler service can be configured programmatically by configuring `GlobalSchedulerService` in the runtime environment:

```
RuntimeEnvironmentBuilder.Factory.get()
    .newDefaultBuilder().entityManagerFactory(emf)
    .knowledgeBase(kbase).schedulerService(globalSchedulerService);
```

Here, the `globalSchedulerService` object is an implementation of `org.jbpm.process.core.timer.GlobalSchedulerService` and the Quartz implementation is `org.jbpm.process.core.timer.impl.QuartzSchedulerService`.

Availability

The availability of an application or system can be viewed as the total amount of time that it provides its services against the total time it is expected to do so. System availability is affected by multiple factors ranging from failure of the system due to hardware/software failures and the known downtime for maintenance and upgrades.

High availability is achieved in applications by having failover mechanisms using which the system can get back to its service provisioning state after a failure. A more optimized system would consider backup mechanisms, which it can immediately switch on to in a failure scenario, thus improving the availability. Scheduled maintenance can be done using a rolling upgrade to ensure high availability. The solutions are usually depicted in the form of deployment architecture, vary according to the software solution, and consider the trade-offs in non-functional requirements.

The following figure depicts the sample deployment architecture that can be applied to the jBPM workbench and runtime, which can cater to high-availability and high-throughput scenarios. The architecture considers a failover mechanism by having a persistent, distributed storage for all data and a load balancer with a passive backup to ensure the switching of nodes upon a partial failure (node failure). jBPM doesn't provide all the components needed in this deployment architecture but has to get third-party software and integrated. We discuss the applicability of these components in the subsequent sections.

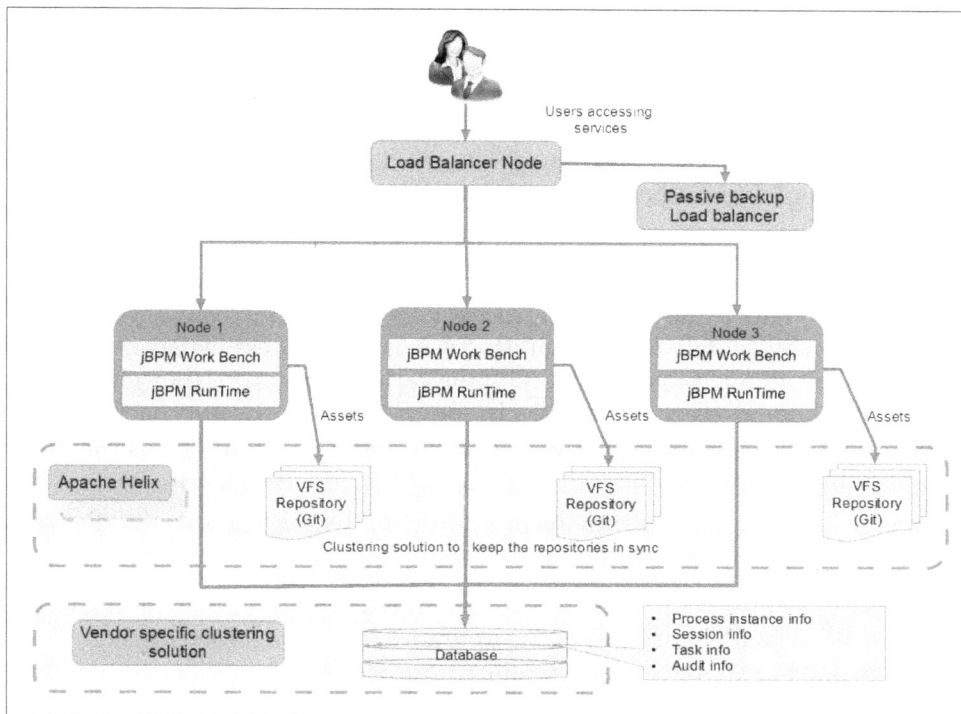

Applicability to workbench

Thinking about the availability of the workspace, we need to consider the following two things:

- Availability of the web-based user interface for modeling
- Availability of the stores where the assets are created during the modeling

The service providing the web-based user interface is hosted on an application server, and the availability means availability provisions provided by the application server. Usually, the provisions are in the form of load balancer-based failover mechanisms, where the load balancer switches the traffic to another node when the node providing the service fails, and the failover mechanism ensures that the node will be back live within a certain timespan to provide the full potential throughput.

Assets are stored in a virtual file system, and by default, it is a Git-based repository. As we have discussed in the *Scaling the business process modeling facility* section, the Git-based repository can be clustered and the assets synchronized. So, even if one node fails, the latest assets will be restored from the other nodes.

Applicability to continuous improvements in processes

From the perspective of the applications developed using the modeling facility, there can be continuous improvements in the application processes. The new version of a process can be deployed to the jBPM runtime in two ways.

- The new version of the process can be deployed as a new process definition, and the old one can be retired by removing it from the knowledge repository so that no new instance of this process version is deployed. However, this approach absolutely doesn't handle the existing process instances. Either they should be allowed to continue with the earlier versions or they should be aborted and reprocessed. The decisions on choosing the approach should be handled case-by-case on the basis of the business scenario that these process definitions are handling.

- Using the process migration facility provided by jBPM. However, the facility is limited to process changes that are non-conflicting.

Thus, process upgrades are not a very smooth process and require careful handling. Further, to achieve availability, either the upgrades have to be done during low-traffic periods or we have to do a rolling upgrade.

Applicability to the process runtime

Availability of the process runtime involves the availability of services to interact with process instances such as process instance life cycles and user interactions. Similar to what we discussed about the workbench facility, there are two things we need to consider:

- Availability of the process instance interactions
- Availability of the process instance data

For the availability of the process instance interactions, we need to have a load balancing mechanism that can switch over the traffic of one node to another upon a failure. Further, we need to be able to persist the process instance data in a nonvolatile storage for a failover. As discussed in the *Scaling the process runtime* section, jBPM supports the persistence of runtime information in relational databases.

Most of the popular relational databases have inbuilt support for availability using clustered storage. This can be utilized to make sure that the persisted process instance data can be made highly available.

Security

Security or application security in this context refers to protecting the services and data provided by jBPM from unauthorized access (authentication) and at the same time ensuring that the users can access the set of services and data authorized for them (authorization).

Another important perspective that we have to consider in a BPM system is providing non-repudiation for all the user interactions. jBPM supports this by providing an audit logging facility for all runtime data changes.

> Non-repudiation assures that a user cannot deny performing an action or operation in the system.

Securing the access of application assets

jBPM, usually deployed in an application server, uses a JEE-compatible standard, that is, **Java Authentication and Authorization Service (JAAS)**, for providing application security. The application server provides this service and a mechanism to register the users.

The default asset repository used to keep the application assets is the Git repository. Further, a Git repository ensures that a change to a repository handled by it (be it an addition of a file or a change of the file) is only allowed for an authorized person.

Authentication is done for the users logging in through the workbench, and for making changes to the repository through Eclipse, the users have to use SSH.

The workbench user management supports the following roles:

- **Admin**: The role that has full access rights. The user who plays the role of admin manages the BPMS.
- **Analyst**: The role that can do the modeling and is associated with a business analyst for creating processes, rules, entities, forms, and so on.
- **Developer**: The role that carries the baton of the process artifacts from the business analyst and develops fully executable code with back-end services and handlers.
- **Business user**: The role that performs operation management by using task management features. The person assigned to this role is the end user of the application, who avails the functionality provided by the application.
- **View only user**: The role that can view the statistics of processes and their performance and is the primary user of the statistics dashboard.

Taking the example of jBPM installed in the JBoss application server (now renamed to WildFly), it provides utilities such as `add-user` to add the users and assign their roles.

Securing the process runtime

The following subsections details the provisions for securing the process runtime, that is, securing the operation management specific operations.

Access security specific to human tasks

The workbench allows only authorized people to access the process runtime capabilities as discussed in the previous section. Apart from the process management access, inside each process definition, each human task is assigned to a user or a role (group).

In an environment using the whole jBPM suite of software, including the process modeler, we can drive this by adding more roles or groups to the system by using the application server capabilities. In an embedded mode, jBPM provides an extension point for implementing an application-specific access security mechanism.

This extension can be done by implementing the `org.kie.api.task.UserGroupCallback` interface and embedding the application-specific logic for validating the authorization of the users to perform tasks. The specific method that we need to implement for attaining this integration is as follows:

```
boolean existsUser(String userId)
```

This interface method is used for determining whether the user attached to a task is valid. For integrating with our application, we can implement this method by using the logic or service used for authentication.

```
boolean existsGroup(String groupId)
```

This interface method is used for resolving whether the group or role attached to a task is valid.

```
List<String> getGroupsForUser(String userId,List<String> groupIds,
List<String> allExistingGroupIds)
```

This is used for resolving the groups (or roles) that are valid for a user.

The `userGroupCallback` interface implementation can be attached to the process runtime by using the `HumanTaskService` factory. The code snippet for doing this is as follows:

```
UserGroupCallBack userGroupCallBack= new CustomUserGroupCallback();
TaskService taskService =
HumanTaskServiceFactory.newTaskServiceConfigurator().
entityManagerFactory(emf).userGroupCallback(userGroupCallBack).
getTaskService();
```

The advantage of this mechanism is that we are not constrained to the user management functionality provided by jBPM but can develop our own. Further, jBPM provides a set of pre-build `userGroupCallBack` functions that can be used in production:

- `org.jbpm.services.task.identity.LDAPUserGroupCallbackImpl`, as the name indicates, can be used for configuring it with your LDAP service.

- `org.jbpm.services.task.identity.JAASUserGroupCallbackImpl`, as the name indicates, can be used for configuring it with your JAAS standard-specific user authentication mechanisms widely used in application server environments. jBPM provides adapters for Oracle WebLogic, IBM WebSphere, and JBoss application servers.

> LDAP (which stands for Lightweight Directory Access Protocol) is an open standard, widely used in small and medium organizations to share user information between services and systems.

Audit logging

In business domains using a BPM, the process defines the business itself. Using these processes, multiple systems and people in the organization interact with one another. In any organization, disputes regarding actions done by people or systems that drive these processes are common. From the perspective of application security, these scenarios are solved using non-repudiation mechanisms, which assure that no user or system can deny these actions. Audit logging is one of the widely used non-repudiation mechanisms, in which every action performed on the system is stored and is later used for resolving a dispute or analyzing the root cause of the dispute. Another advantage is that we can use this data to analyze and find out the performance and quality indicators of the business processes.

An audit log helps us to retrieve information about what happened to a process instance, when it happened, and who triggered it.

jBPM provides a generic audit logging mechanism that comprehensively covers the life cycle of a business process. The audit log is stored as three data models:

- **Process instance log**: Stores the data corresponding to interactions with the process instance life cycle, for example, starting of a process instance, stopping of a process instance, or aborting a process instance. Using the attributes of the instance log, we can trace back the process definition, process version, process instance, user identity, and so on, which are associated with a life cycle change.

- **Node instance log**: Stores the data corresponding to the life cycle of a node in the process. A node refers usually to an activity in the business process. The attributes of this data help us to trace back to the process definition, process version, process instance, user identity, time, and workitem on which this incident occurred.

- **Variable instance log**: Stores the data corresponding to the changes in process variables in a process instance.

The following table lists the data available in the audit log data model:

Process instance log		
Field	**Type**	**Description**
ID	BIGINT(20)	Identity of the log table
Duration	BIGINT(20)	Lifetime of the process instance when the incident occurred
End date	DATETIME	Represents the time when the process instance ended, applicable only if the process instance is stopped or aborted
External ID	VARCHAR(255)	An external ID provided for the identification of a process instance from the domain data
User identity	VARCHAR(255)	Identity of the user who initiated the process instance.
Outcome	VARCHAR(255)	Outcome of the process information, primarily used to store information such as error code, in case the process stops due to an error event
Parent process instance ID	BIGINT(20)	The identifier of the parent process instance
Process ID	VARCHAR(255)	Identifier of the process definition
Process instance ID	BIGINT(20)	Unique identifier of the process instance
Process name	VARCHAR(255)	Name of the process definition

Process version	VARCHAR(255)	Version of the process definition
Start date	DATETIME	Date on which the process instance was started
Status	INT(11)	

Status	Integer value	Description
ACTIVE	1	Represents a live process instance
COMPLETED	2	Represents a completed process instance
ABORTED	3	Represents an aborted process instance
SUSPENDED	4	Represents a suspended process instance

The preceding table provides the possible values of this field and what they mean

Node instance log

Field	Type	Description
ID	BIGINT(20)	Unique identifier
Connection	VARCHAR(255)	Identifier of the sequence flow that led to this node instance
Log date	DATETIME	Date at which the node was triggered
External ID	VARCHAR(255)	External identifier associated with the process instance
Node instance ID	VARCHAR(255)	Identifier of the node instance
Node name	VARCHAR(255)	Name of the node from the process definition
Node type	BIGINT(20)	The type of node or activity, for example, service task
Process ID	VARCHAR(255)	Identifier of the process definition that this node is a part of
Process Instance ID	BIGINT(20)	Identifier of the process instance that this node is a part of
Type	INT(11)	Indicates whether the log was updated on entry or exit
Workitem ID	BIGINT(20)	Identifier of the workitem that this node refers to

Variable instance log

Field	Type	Description
ID	BIGINT(20)	Unique identifier
Log date	DATETIME	Time at which the change in this variable occurred
External ID	VARCHAR(255)	External identifier associated with the process instance
Old value	VARCHAR(255)	Previous value of the variable
Process ID	VARCHAR(255)	Process ID of the definition

Process instance ID	BIGINT(20)	Process instance identifier
Value	VARCHAR(255)	Current value of the variable
Variable ID	VARCHAR(255)	Identifier, variable name
Variable instance ID	VARCHAR(255)	Additional information when a variable is defined on the composite node level to distinguish between top-level and embedded-level variables

Apart from its use for security, this log information can be analyzed to find out various performance indicators of the process and the organization. The dashboard builder can be used to build reports from these logs.

Maintainability

The maintainability of a system can be considered to be a measure to determine how easily the repair actions can be performed. When we say repair, we need to discuss the following:

- The ease of fixing issues in a deployed system (if any)
- Improvements in the system to match the changing business needs
- Coping with infrastructure changes in the deployment environment

In the system of our consideration, a BPM, changes in business logic are more frequent. So, one of the main factors from a maintainability perspective is the ease of improving the process executable. This is one area that jBPM excels in; as we have already discussed in earlier chapters, jBPM provides a full-fledged modeling, simulation, and deployment tooling environment. The actors, from this perspective, business analysts and developers, can use the tooling to model, simulate, test, and deploy the process changes.

Another aspect is the infrastructure or the environment in which jBPM is deployed and maintained in production. jBPM supports multiple deployment architectures as discussed in *Chapter 8, Integrating jBPM with Enterprise Architecture*, and by default, it focuses on deployment in a JEE environment, where it is deployed inside a JEE application container, with persistent data storage in a conventional relational database.

The architecture of the system is based on the following standards:

* Modeling based on BPMN
* Simulation based on BPsim
* Human tasks based on WS-HT
* Persistence based on JPA
* Transaction management based on JTA

The advantage is that jBPM easily fits into our current production environment, and as the environment evolves, so does jBPM with its development community playing an active role in enterprise middleware architecture. Compliance to the standards and modularity of the system ensures that our client doesn't fall into a vendor lock-in scenario, with parts of the system being easily replaceable.

In the previous chapters, we have already explained the "how" of the functionalities discussed in this section.

Fault tolerance

Fault tolerance indicates the ability to operate in a predictive manner, when one or more failures happen in the system. In Java-based applications, these faults are managed using exception handling mechanisms. jBPM is no exception; it uses the exception handling approach to be fault-tolerant.

Exception handling in process definitions

We can specify the occurrence and the handling mechanisms that happen in a business process using BPMN elements, as follows:

* Error events can be used to specify the occurrence of an unexpected situation. Compared to Java programming, this is similar to throwing an error.
* Compensation can be used to specify what to do when an error has occurred; this is similar to the catch operation construct in a Java program.

The advantage of using exception handling at a process level is that the exception scenarios are visible in the process, thus making the monitoring and analysis of these scenarios easier, thereby contributing to continuous improvements of the process.

Exception handling for domain-specific processes

When we define new custom workitem handlers that form custom, business-specific activities in a process, we can specify mechanisms to handle exception scenarios. jBPM by default provides the following decorators for handling an exception scenario in an activity:

- `SignallingTaskHandlerDecorator`: This decorator catches an exception during the life cycle methods of an activity and signals the process instance using a configurable event. These events can be caught in the process definition, and subsequent actions can be taken. This decorator can be specified while registering the workitem handler to a session.

  For example:

  ```
  String eventType = "Mail-Service-failed";
  SignallingTaskHandlerDecorator signallingTaskWrapper = new
  SignallingTaskHandlerDecorator
  (MailServiceHandler.class, eventType); signallingTaskWrapper.
  setWorkItemExceptionParameterName
  (ExceptionService.exceptionParameterName);
  ksession.getWorkItemManager().registerWorkItemHandler
  ("Mail Task", signallingTaskWrapper);
  ```

  In this example, we register a handler for sending mails by using `MailServiceHandler.class`, and during exception scenarios, the `"Mail-Service-Failed"` event is signaled to the process instance.

- `LoggingTaskHandlerDcorator`: This decorator catches the exceptions during the life cycle methods of the activity of the logging mechanism. This feature can be used in less critical areas where a process exception can just be a warning in the log.

Summary

In this chapter, we discussed the non-functional characteristics of jBPM that are critical in building a production-ready application based on BPM. Further, we discussed sample deployment architectures that are possible with jBPM to meet various requirements and configurations, and customization available to include certain characteristics in the jBPM system.

A
The Future

As we reach the conclusion of *Mastering jBPM6*, traversing through the why, what, and how of jBPM, it becomes obligatory to discuss what the future holds. Of course, we are not foretellers, but when it comes to a technology that we use at the heart of our enterprise architecture and evangelize, we need to understand the place of this system in future computing.

The computing world is going through a brisk evolution, automation is the key and **Internet of Things (IoT)** is at the door, revolutionizing the world as we know it. In this short Appendix, we discuss the place of jBPM and the related technologies in the future, by detailing the following trends in computing:

- Business programming
- Convergence of enterprise architectures

Business programming

Enterprise applications are driven by business needs. In other words, we program these applications to satisfy certain business requirements. Further, in a traditional software development life cycle, the requirements are communicated by the business users to the software developers, who consolidate and convert them into executable implementations.

This conventional way of developing applications using programming languages has already been superseded by business-level execution languages that talk business (or use business-friendly jargon) such as BPEL. In the current arena, to improve usability and maintainability, they are evolving into applications supporting visual business-oriented programming of which jBPM is a flag bearer.

"A picture is worth a thousand words"—this quote can explain why we need visual programming. The objective is obviously to communicate effectively. No longer is programming considered to be wizardry, programs and their logic are part of business operation and should be easily available to people across the organization.

Further, we can confidently say that this is where programming is being led. The users must be able to view how the operations are done, how decisions are made, and what the impacts of a change are. jBPM modeling facilities are a huge leap into a completely visual programming experience. Currently, the jBPM arsenal allows us to model business processes and business rules. We can model data and user interfaces through a limited but useful functionality, which minimizes the development and deployment costs. We can also simulate a business process and see how it works along with analyzing the related performance metrics.

Already trending are buzzwords such as IoT or computing everywhere, from the perspective of BPM, these are activities in business operations. Visual programming can help us to encapsulate the logic for communicating with these pervasive devices and give the business, a picture of everyone connected to the process and everyone operating it.

Convergence of enterprise architectures for real-time and predictive analytics

Gone are the days when the IT infrastructure contained separate applications for operations, analytics, and administration. Today's IT infrastructure needs enterprise application suites that can deliver operations, analytics, and administration together, and can collaborate to share information (in real-time), which makes their functioning more efficient. For example, data from operations is used by analytics to detect fraudulent operations, thereby making operations more secure. Further, the future asks for more seamless interactions between these operations.

From a technical perspective, we can see these as a merger of multiple design patterns. We can see systems that efficiently merge **service-oriented architecture (SOA)** and event-driven architecture. For example, the LAMBDA architecture merges batching, analytics, and service provisioning.

Such convergence can lead to more effective technologies such as predictive analytics. A BPM-based system can be used to produce, consume events, and interact with services. Thus, predictive analytics on these events is an obvious improvement. Moreover, BPMs can be used to take action on the results (or decisions) obtained by using these predictive analytic engines. This can have foreseeable applications in the following domains:

- Fraud detection.
- Suggestive marketing.
- More resilient manufacturing automation, and so on.

The BLIP, that we have already discussed in *Chapter 1, Business Process Modeling – Bridging Business and Technology*, is certainly moving to this goal. It provides a way to integrate business logic spread across operations and enable them to work collaboratively to achieve efficient business operations; jBPM is a vital part of BLIP.

B

jBPM BPMN Constructs Reference

This Appendix is a quick reference for the BPMN constructs supported by jBPM. For a complete BPMN 2.0 guide, please refer to the Object Management Group specifications and guides available at http://www.bpmn.org. You can also refer to *Chapter 5, BPMN Constructs,* in this book for an in-depth, sample-driven discussion of jBPM constructs. For easy indexing and searching, this reference describes the constructs as displayed and grouped within the KIE process definition editor.

Tasks

A task represents an action that needs to be performed.

User

- **BPMN element**: <bpmn2:userTask>.
- **Description**: Task that requires a human interaction with a UI or a programmed API.
- **Configuration**: Use Actors or Groups for assigning the task to the users. Use the Assignments, DataInputSet, and DataOutputSet properties for task parameter mapping to the enclosing process instance.

Send

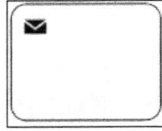

- **BPMN element**: `<bpmn2:sendTask>`
- **Description**: General-purpose task for a *send* message action
- **Configuration**: The `MessageRef` property is a key to the ID attribute of a `message` element (`bpmn2:message`), which must be defined in the process scope

It needs a custom WorkItemHandler to be registered with the *send* key.

Receive

- **BPMN element**: `<bpmn2:receiveTask>`
- **Description**: General-purpose task for a *receive* message action
- **Configuration**: The `MessageRef` property is a key to the ID attribute of a `message` element (`bpmn2:message`), which must be defined in the process scope

It needs a custom WorkItemHandler to be registered with the *receive* key.

Manual

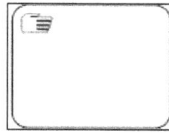

- **BPMN element**: `<bpmn2:manualTask>`
- **Description**: Task whose purpose is oriented to documenting that an action must be performed manually and that it can be ignored by the engine

Service

- **BPMN element**: `<bpmn2:serviceTask>`.
- **Description**: Service task that supports a Java or SOAP WebService call.
- **Configuration**: Use the `ServiceImplementation`, `ServiceInterface`, and `ServiceOperation` properties to configure the callable service/class. The WebService implementation requires the WSDL URL to be imported in the global scope (process imports).

Business rule

- **BPMN element**: `<bpmn2:businessRuleTask>`
- **Description**: Task that executes a Drool rule
- **Configuration**: Use the `Ruleflow Group` property to select the name of the Rule Group that defines the set of rules that you need to execute

Script

- **BPMN element**: `<bpmn2:scriptTask>`
- **Description**: Task that executes Java or MVEL scripts
- **Configuration**: Use the `On Entry Actions`, `On Exit Actions`, and `Script language` properties

None

- **BPMN element**: `<bpmn2:task>`
- **Description**: Ad hoc task
- **Configuration**: You must register a WorkItemHandler with the *task* key

Subprocesses

The subprocess represents a group of tasks which act together to perform a part of the process.

Reusable

- **BPMN element**: `<bpmn2:callActivity>`.
- **Description**: Lets you call a process whose definition resides outside the current process.

- **Configuration**: Use `CalledElement`, `Independent`, and `WaitForCompletion` to configure which existing process definition to call and how to call it: as part of the calling process or as a new process instance (`Independent`), in a synchronous or asynchronous fashion (`WaitForCompletion`). Use `Assignments`, `DataInputSet`, and `DataOutputSet` to map variables from/to the calling process.

Multiple instances

- **BPMN element**: `<bpmn2:subProcess>`
- **Description**: Lets you loop (create multiple instances of a group of elements)
- **Configuration**: Use `CollectionExpression`, `Variable Definitions`, and `Variable Name` to configure the loop and assign the variable to pass inside the loop

Ad hoc

- **BPMN element**: `<bpmn2:adHocSubProcess>`
- **Description**: Lets you define an unstructured subprocess definition
- **Configuration**: The `AdHoc ordering` attribute tells the engine to execute a multi-instance subprocess in parallel or sequentially

Embedded

- **BPMN element**: `<bpmn2:subProcess>`
- **Description**: Lets you define an embedded process definition (not reusable from other process definitions)
- **Configuration**: `Variable definitions` lets you configure variables at the subprocess scope

Events

- **BPMN element**: `<bpmn2:subProcess>`
- **Description**: Lets you define an embedded subprocess that can be triggered by a specific event (for example, `Signal`) and is executed in an asynchronous fashion
- **Configuration**: No specific configuration required

Start events

- **BPMN element**: `<bpmn2:startEvent>` and a child element that defines the event type as shown in the following image:

- **Description**: Acts as a process trigger and can only be a catching event.

The supported start events are as follows (see the preceding image, left to right):

- **None**
- **Message**: <bpmn2:messageEventDefinition>
- **Timer**: <bpmn2:timerEventDefinition>
- **Escalation**: <bpmn2:escalationEventDefinition>
- **Conditional**: <bpmn2:conditionalEventDefinition>
- **Error**: <bpmn2:errorEventDefinition>
- **Compensation**: <bpmn2:compensationEventDefinition>
- **Signal**: <bpmn2:signalEventDefinition>

End events

- **BPMN element**: <bpmn2:endEvent> and a child element that defines the event type as shown in the following image:

- **Description**: Acts as a process trigger and can only be a throwing event.

The supported end events are as follows (see the preceding image, left to right):

- **None**
- **Message**
- **Escalation**
- **Error**
- **Cancel**: <bpmn2:cancelEventDefinition>
- **Compensation**
- **Signal**
- **Terminate**: <bpmn2:terminateEventDefinition>

Catching intermediate events

- **BPMN element**: `<bpmn2:intermediateCatchEvent>` and a child element that defines the event type (see the previous events list)
- **Description**: Events that catch the triggering of a matching throwing event

The supported events are as follows (see the preceding image, left to right):

- **Message**: Can be a boundary event
- **Timer**: Can be a boundary event
- **Escalation**: Can be a boundary event
- **Conditional**: Can be a boundary event
- **Error**: Can be a boundary event
- **Compensation**: Can be a boundary event
- **Signal**: Can be a boundary event

Throwing intermediate events

- **BPMN element**: `<bpmn2:intermediateThrowEvent>`
- **Description**: Events that throw event triggering

The supported events are as follows (see the preceding image, left to right):

- Message
- Escalation
- Signal

Gateways
Gateways control how the process flows.

Data-based exclusive (XOR)

- **BPMN element**: `<bpmn2:exclusiveGateway>`
- **Description**: Used to choose alternative sequence flows

Event-based gateway

- **BPMN element**: `<bpmn2:eventBasedGateway>`
- **Description**: Used to trigger sequence flows upon the occurrence of some event

Parallel

- **BPMN element**: `<bpmn2:parallelGateway>`
- **Description**: Used to create parallel sequence flows where all paths are evaluated

Inclusive

- **BPMN element**: `<bpmn2:inclusiveGateway>`
- **Description**: Used for creating alternative flows where all paths are evaluated

Data objects

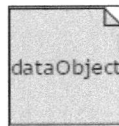

- **BPMN element**: `<bpmn2:dataObject>`
- **Description**: Data objects show the reader which data is required or produced in an activity

Swimlanes

Swimlanes represent grouping of process actors or roles with respect to the process tasks.

Lane

- **BPMN element**: `<bpmn2:lane>`
- **Description**: Used for organizing activities within a group according to a user or a user group

Artifacts

Artifacts are elements useful for documentation purposes.

Group

- **BPMN element**: `<bpmn2:group>`
- **Description**: The group visually arranges different activities together; it does not affect the flow in the diagram

Annotation

- **BPMN element**: `<bpmn2:annotation>`
- **Description**: The annotation is used for giving the reader of the diagram an understandable comment/description

Index

audit data, generating 117, 118
audit service, tips 115
business query, writing 118
data provider, adding 119
default history logs 115, 116
examples 114
KPI panels 120, 121
new dashboard page, creating 120, 121
notes 122
ProcessBAM unit test 117, 118
URL 113
using, with Dashbuilder 116
banking
with BPM 13
Bean Managed Transactions (BMT) 203
Bitronix
using, for local transaction 204
BLIP
about 24
Business Processes 25
Business Rule 25
Drools Expert 26
Drools Fusion 26
Drools Guvnor 26
Event Stream Processing 25
features 24
boundary events 137
build process, for managed repository projects
Build & Deploy 91
Build & Install 90
Business Activity Monitoring. *See* **BAM**
Business Logic Integration Platform.
See **BLIP**
business process deployment
knowledge base, creating 46
runtime engine, creating 47
runtime manager, creating 46
starting 47
Business Process Diagram (BPD) 6
Business Process Execution Language (BPEL) 6
Business Process Management (BPM)
about 2, 3
banking 13
financial services 13

supply chain management 11, 12
usage 10, 11
Business Process Management System (BPMS) 2
Business Process Model and Notation (BPMN)
about 6
conformance 7
core elements 7
URL 6
versus BPEL 6
business process modeling
design patterns 14
facility, scaling 252-254
business process simulation 5
Business Process Simulation Interchange Standard (BPSim) 6
business programming 271, 272
business solution tools
integrating 27, 28

C

CommandExecutor interface
about 192, 193
bath execution 193, 194
Drools Rules 192
Process/Task 192
Runtime 192
compensation
about 148
activity 149
Boundary Compensation event 149
Intermediate Compensation event 149
triggering, with signals 150
Complex Event Processing (CEP) 114
complex gateway 135
conditional branching
about 130
complex gateway 135
Drools 131
event-based gateway 133, 134
exclusive (XOR) gateway 131
inclusive (OR) gateway 132, 133
conformance, BPMN
BPEL process execution conformance 7

G

gateway direction property
converging 129
diverging 129
mixed 129
unspecified 129
gateways
about 8, 129, 283
concepts 129
conditional branching 130
data-based exclusive (XOR) 283
direction property, defining 129
event-based gateway 283
inclusive 284
parallel (AND) gateway 130
parallel 283
getKieClasspathContainer() function 173
Git
assets, managing 86
changes, making 84
committing 84
KIE workbench project editor 93
new repository 86
overview 82
pushing, to remote repository 85
remote repository, cloning 83
working with 82
globals, KieSession 126, 191, 192
governance workflow
about 87
asset promotion 87
project build 87
release 87
repository creation or configuration 87
group, artifacts 9
GroupId-ArtifactId-Version (GAV) 173

H

Hibernate
URL 31
Human Interaction Management (HIM)
about 3, 4
automatic activities 3
manual activities 3
human task activities, roles
actual owner 154

business administrator 154
excluded owners 154
potential owners 154
task initiator 154
human task service
about 196
admin service 198
attachment service 198
CommandService 196
content service 198
deadline service 198
DefaultUserInfo instance 196
EntityManagerFactory instance 196
instance service 198
interceptors 198
LocalTaskServiceFactor 196
notification service 198, 199
query service 198
register task event listeners 196
TaskCommand 197
TaskContext 197
TaskDeadlinesService 196
TaskFluent class 199, 200
transactions 198
UserGroupCallback 196, 197

I

implicit termination pattern 17, 18
inclusive (OR) gateway 133
interface methods, workitem handler
org.kie.internal.runtime.Cacheable.
close() 210
org.kie.internal.runtime.Closeable.
close() 210
intermediate events
catching 282
throwing 282
Internet of Things (IoT) 271

J

Java Authentication and Authorization
Service (JAAS) 263
Java Content Repository (JCR) 98
Java Message Service (JMS) 23
Java Naming and Directory
Interface (JNDI) 198

[PACKT] open source*
PUBLISHING   community experience distilled

Thank you for buying
Mastering jBPM6

About Packt Publishing

Packt, pronounced 'packed', published its first book, *Mastering phpMyAdmin for Effective MySQL Management*, in April 2004, and subsequently continued to specialize in publishing highly focused books on specific technologies and solutions.

Our books and publications share the experiences of your fellow IT professionals in adapting and customizing today's systems, applications, and frameworks. Our solution-based books give you the knowledge and power to customize the software and technologies you're using to get the job done. Packt books are more specific and less general than the IT books you have seen in the past. Our unique business model allows us to bring you more focused information, giving you more of what you need to know, and less of what you don't.

Packt is a modern yet unique publishing company that focuses on producing quality, cutting-edge books for communities of developers, administrators, and newbies alike. For more information, please visit our website at www.packtpub.com.

About Packt Open Source

In 2010, Packt launched two new brands, Packt Open Source and Packt Enterprise, in order to continue its focus on specialization. This book is part of the Packt Open Source brand, home to books published on software built around open source licenses, and offering information to anybody from advanced developers to budding web designers. The Open Source brand also runs Packt's Open Source Royalty Scheme, by which Packt gives a royalty to each open source project about whose software a book is sold.

Writing for Packt

We welcome all inquiries from people who are interested in authoring. Book proposals should be sent to author@packtpub.com. If your book idea is still at an early stage and you would like to discuss it first before writing a formal book proposal, then please contact us; one of our commissioning editors will get in touch with you.

We're not just looking for published authors; if you have strong technical skills but no writing experience, our experienced editors can help you develop a writing career, or simply get some additional reward for your expertise.

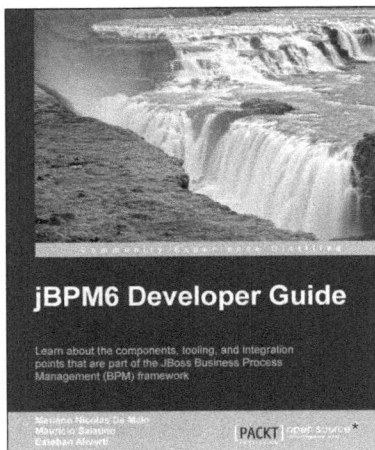

jBPM6 Developer Guide

ISBN: 978-1-78328-661-4   Paperback: 310 pages

Learn about the components, tooling, and integration points that are part of the JBoss Business Process Management (BPM) framework

1. Model and implement different business processes using the BPMN2 standard notation.

2. Understand how and when to use the different tools provided by the JBoss Business Process Management (BPM) platform.

3. Learn how to model complex business scenarios and environments through a step-by-step approach.

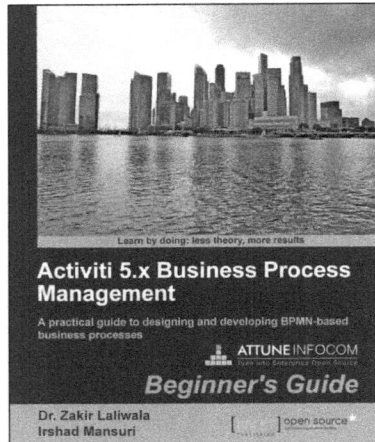

Activiti 5.x Business Process Management Beginner's Guide

ISBN: 978-1-84951-706-5   Paperback: 276 pages

A practical guide to designing and developing BPMN-based business processes

1. Detailed coverage of the various BPM notations used for business process development.

2. Learn how to implement business processes based on real world examples.

3. Understand how to deploy workflows using process engine APIs.

4. Create Advance workflows using BPM Notations.

Please check **www.PacktPub.com** for information on our titles

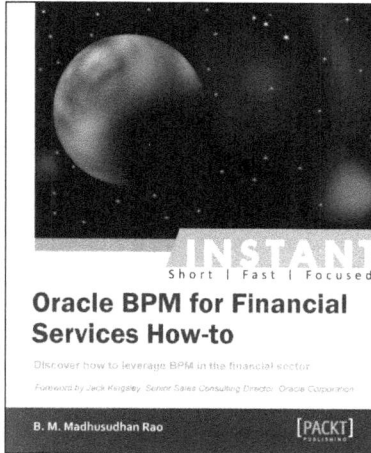

Instant Oracle BPM for Financial Services How-to

ISBN: 978-1-78217-014-3   Paperback: 62 pages

Discover how to leverage BPM in the financial sector

1. Learn something new in an Instant! A short, fast, focused guide delivering immediate results.

2. Simplifies complex business problems for financial services.

3. Optimize, enhance, and modify your business processes.

4. Includes enterprise architecture best practices.

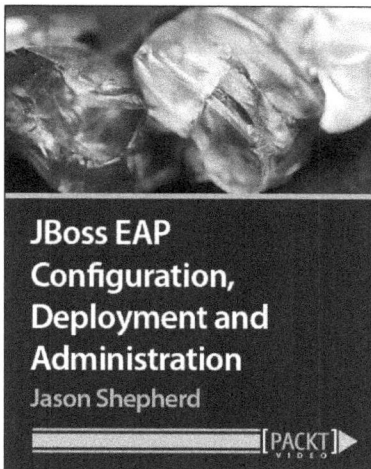

JBoss EAP Configuration, Deployment, and Administration [Video]

ISBN: 978-1-78216-248-3   Duration: 02:08 hours

Detailed demonstrations to help you harness one of the world's top open source JEE projects

1. Learn about everything from installation, configuration, and debugging to securing Java EE applications — ideal for JBoss application developers.

2. In-depth explanations of JBoss EAP features, and diagrams to help explain JBoss and Java internals.

3. Covers everything from JBoss EAP essentials to more advanced topics through easy-to-understand practical demonstrations.

Please check **www.PacktPub.com** for information on our titles

www.ingramcontent.com/pod-product-compliance
Lightning Source LLC
Chambersburg PA
CBHW080926220326
41598CB00034B/5698